Telecosmos

Telecosmos

The Next Great Telecom Revolution

John Edwards

A JOHN WILEY & SONS, INC. PUBLICATION

Library of Congress Cataloging-in-Publication Data:
Edwards, John.
   Telecosmos : the next great telecom revolution / John Edwards.
     p. cm.
   Includes bibliographical references and index.
   ISBN 0-471-65533-3 (pbk.)
   1. Telecommunication—Technological innovations.   I. Title.

TK 5101.E33 2005
621.382—dc22

10 9 8 7 6 5 4 3 2 1

*To Jonathan M. Bird, radio enthusiast and online pioneer,
who welcomed me into the world of telecommunications.*

Canst thou send lightnings, that they may go, and say unto thee, Here we are?
Job 38:35

Contents

List of Figures

Introduction

I started exploring telecommunications' frontiers at the tender age of nine, way back in 1964. That was the year I visited the New York World's Fair and found myself, quite unexpectedly, drafted into a corporate public relations demonstration.

Back then, nearly all U.S. telecommunications—hardware, software and service—was concentrated in the hands of a giant monopoly—the American Telephone and Telegraph Co. AT&T's showplace at the fair was the Bell System Pavilion. (Fig. Intro.-1). The Bell System, for those too young to remember, was AT&T's conglomeration of regional telephone operating companies. A federal court order, issued 20 years after the fair closed, forced AT&T to divest itself of the firms.

Like an ancient cathedral, the pavilion was designed to be simultaneously functional and awe inspiring. (No surprise, since in the 1960s AT&T had nearly as much raw political power as the medieval church.) Situated on 2.5 acres of prime, reclaimed Queens swampland, the pavilion's upper section was a massive "floating wing," measuring 400 feet long, 200 feet wide, and 87 feet high. Held aloft by a set of four 24-foot-tall pylons, the gleaming white structure (marred only by traditional blue Bell System logos on each side) looked as though it were poised to take off, soar over the nearby Pool of Industry, and perhaps buzz the hapless New York Mets baseball team playing at nearby Shea Stadium. Altogether, the building required 7,250 cubic yards of concrete, 900 tons of reinforcing steel, 3,000 tons of structural steel, and 450 plastic reinforced fiberglass panels.

Most of the floating wing's 41,000 square feet of useable space was consumed by a nonthrill ride that people waited as long as several hours to expe-

Figure Intro.-1   *The Bell System Pavilion, AT&T's showplace at the New York World's Fair in 1964.*

rience. As visitors sat in "floating armchairs," a conveyor belt tugged them along at a pace of 70 feet per minute past animated exhibits tracing the history of communications (including, naturally, Dr. Bell and Mr. Watson). Beneath the floating wing, AT&T constructed a 40,000-square-foot exhibit hall that housed technology displays, live demonstrations, and even audience participation games. A 140-foot tower in front of the pavilion (the only possible obstruction to a smooth take-off) held a one-ton microwave horn antenna that was designed to relay video from the fair to a receiver located on top of Manhattan's Pan Am Building (now the MetLife Building).

The Bell System Pavilion didn't last long. In its predivestiture heyday, AT&T and its financial might could afford to have the massive structure demolished shortly after the fair's end. In all, the pavilion was open to the public for less than two years, from April through October in both 1964 and 1965.

INFORMATION PORTAL

By plan, the Bell System Pavilion was designed to serve as an information portal to the Third Great Telecom Revolution. (The first two telecom revolutions were launched in the mid- and late-19th century by the creation of, respectively, the global telegraph and telephone networks.) The exhibits included a full-scale replica of the first Telstar communications satellite, a presentation on microwave links, a selection of stylish new desktop Touch Tone telephones, and six interconnected AT&T PicturePhones that visitors could examine and play with.

Whereas some of the presented concepts never took hold, or only became available much later than forecast and in greatly changed forms, the exhibits did much to enlighten a public that was still accustomed to using black rotary-dial telephones. Conspicuously absent from the exhibit hall, however, was an array of communications technologies that would become commonplace by the early 21st century, including such innovations as mobile phones, PDAs, wireless hotspots, global positioning system (GPS) technology, and fiber-optic cables. That's a lot to miss. But then, it's easy to nitpick from a 21st century vantage point.

BACK TO ME

My role in AT&T's great telecom public relations campaign took place in the summer of 1964. While strolling along the pavilion's exhibit hall with my parents and little brother, we stopped to see a PicturePhone demonstration. After spending several minutes viewing an earnest presentation on the technology and its potential, the hostess—a young woman—looked at me and smiled. (I must have looked particularly cute that day.) She asked me whether I wanted to participate in a test. As a youngster the word "test" carried certain, strongly negative connotations. But before I had a chance to answer, I found myself being hustled toward a small stage that held a desk, a chair, and a PicturePhone prototype. On the PicturePhone's screen I could see the smudgy black-and-white image an old lady wearing a hat that looked something like an upside down flower pot.

On closer examination, the old lady appeared confused. A stream of sounds fluttered from her mouth: "Umph. Anxth. Hello? Hello? Hello?" She paused for a second or two, looked anxiously around, and resumed talking. "Hello? Hello? Hello?"

The Bell hostess plopped me into the chair. As I sat down, the confusion quickly fell off the lady's face and was replaced with a smile. "I now see someone," she said, looking directly at me from the screen. "It's a little boy. Hello, young man."

Since I was the only kid in the immediate vicinity, I quickly figured out that she was talking to me. "Hi," I said, uncharacteristically at a loss for words. "That's a nice hat you have."

The old lady tossed aside my well-intentioned complement. "Oh, you're such a cute little boy," she replied. "Are you enjoying the fair?"

"Yeah," I replied. I paused awkwardly for a few seconds. Again at a loss for words, I decided to repeat her question. "Are *you* enjoying the fair?" I asked.

"Oh, dear me, no. I'm not at the fair, son," she answered. "I'm visiting Disneyland."

Her words hit my ears and rebounded inside my soft head like a thunder. Disneyland! California! The fabled home of Mickey, Minnie, and Tomorrowland! Even at the tender age of eight, I knew that Disneyland was on the other side of the country. (Actually, my father had taught me that valuable geographic lesson a couple years earlier, shortly after Walt's Sunday evening TV program inspired me to launch a brief, yet spirited, "let's all go to Disneyland" lobbying campaign.) Talking to Disneyland! At our home, talking on the phone to my Aunt Hannie, who lived on the other side of Queens, was considered a big deal. Yet here I was, not only chatting with—but actually seeing—a person thousands of miles away. This was the big time!

Before I had a chance to fully appreciate the moment, or even ask the old lady to say hi to Mickey for me, the hostess whisked me out of the chair and sent me back to into the realm of primitive mortals (namely, those who have never used a PicturePhone). The hostess resumed her patter. "The Bell System created this transcontinental PicturePhone hook-up to demonstrate both the practical and human potential of this ground-breaking technology," she said in a sing-song, rote-style of voice. "In just a few years, people around the world will not only hear but see . . . blah, blah, blah. Grandparents will be able to see their grandchildren . . . blah, blah, blah. Paving the path to worldwide peace . . . blah, blah, blah." I didn't listen very carefully; I was already completely sold on the technology. Slick marketing had claimed another victim.

PicturePhone was never out of my mind over the next few months. I could hardly wait for the day when the green and white New York Telephone truck would pull up in front of our home and the technician would expertly install a PicturePhone on the family's official bill-paying desk. Yeah, right. It would be a long and fruitless wait, of course. Even Touch-Tone dialing, another emerging technology aggressively touted by AT&T at the fair, didn't arrive in our home until 1976. I wouldn't have another PicturePhone-like experience until 2002, when I installed a Webcam on my PC and had a five-minute chat (in color!) with my brother. I haven't used the technology since.

I, TELECOM JUNKIE

Although the Bell System Pavilion entered history before I reached my 11th birthday, the structure and the exhibits it housed made a profound and everlasting impact on me. Ma Bell's fancy showcase awakened within me an interest in—and a love of—telecommunications that I have carried with me throughout my life. The idea of sending human intelligence through wires, or the air, appeals to me in a way that I cannot adequately explain.

In September 1966, a year after the fair and after the pavilion closed forever, I met Jonathan Bird, who helped me get my first hands-on experiences with telecommunications technologies. Jon, a year-and-a-half older than me, was a radio amateur—a ham. We became close friends (best friends, really), and he introduced me to the glorious universe of electronics, transmitters, receivers, and radio propagation. That's why I've dedicated this book to Jon, WA2MJK. If he were alive today, I'm sure he would get one hell of a kick out of pocket-sized mobile phones, the Internet, GPS, satellite TV, and all of the other fantastic technologies that are now an integral part of daily life.

Now, even more exciting telecom technologies are on the way. Although the first years of the 21st century have been tough for the telecom industry—with bankruptcies, corporate scandals, and falling stock prices filling the headlines—research hasn't stopped pushing forward. In laboratories world-wide, fundamental discoveries are being made and sophisticated new technologies are being developed that will shape tomorrow's telecom world, making life easier and bringing people closer together. Many of these emerging technologies, like Voice over Internet Protocol (VoIP), radio-frequency identification (RFID), wireless networks, and Web videoconferencing (the PicturePhone's direct descendant), have the potential to become tremendous, society-altering forces.

So turn the pages and get an advance peek at the many different telecom concepts that will become available to both consumers and businesses in the years ahead. Think of this book as your portable Bell Systems Pavilion—but without the long waiting lines and silly demonstrations.

JOHN EDWARDS

Chapter *1*

On the Menu— Telecom Services

Telecommunications has made rapid strides over the past quarter century. We've gone from a limited selection of desktop phones, controlled by a government-sanctioned monopoly, to a virtually endless choice of wired, wireless and Internet-based communications services. Although many people complain about today's chaotic telecom market, they probably don't remember the limited, high-priced communications options that were available before the telecom boom of the 1980s and 1990s.

Telecom will continue to advance at a furious pace over the next couple of decades. Even the humble home telephone, a mainstay since the days of Alexander Graham Bell, will likely disappear, perhaps replaced by an Internet-based communications appliance or by an Internet-connected mobile phone. (This is a trend that may already be happening in light of the fact that the number of U.S. residential phone lines has been falling since 2001.) Ultimately, people will probably have one phone and one phone number that they'll use both at home and on the road (and they'll take it with them wherever they travel in the world).

In many respects, today's telephone service is little changed from the technology our grandparents used. Despite advancements on numerous technological fronts, as well as the widespread use of mobile phones, most people still think of telephones as those familiar little devices that sit on desks, nightstands, and end tables.

Telecosmos: The Next Great Telecom Revolution, edited by John Edwards
ISBN 0-471-65533-3   Copyright © 2005 by John Wiley & Sons, Inc.

This situation will soon change. Over the next few years, several develop-ments will combine to make telephone service more innovative, less expen-sive, and increasingly useful. Although conventional telephones may linger on for a decade or more, people will increasingly rely on alternative telecommu-nication modes to keep in touch with friends, family, and business contacts.

1.1 END OF THE LINE FOR WIRELINE?

The U.S. carrier market is facing a tough and uncertain future, as total wire-line service revenues continue to dwindle. In-Stat/MDR, a technology market research firm located in Scottsdale, Arizona, finds that long-distance service providers are the group facing the greatest challenge, as their core revenue sources—voice and long-haul private lines—show the greatest revenue declines. Long-distance providers also have very little market share in the few growing consumer telecom services, most importantly broadband.

"Over time, as wireless continues to mature and becomes seamless and reli-able, the need to put up new wired infrastructure will decrease to the point of no longer needing it," predicts John Bartucci, senior director of product man-agement for Telular, a wireless equipment manufacturer located in Vernon Hills, Illinois. "It's a question of putting up poles and stringing wires, or digging trenches to lay cables, versus putting up wireless towers. Assuming there are no health risks associated with all the wireless stuff we've got floating about, I believe that we could see the end for the need for wired technologies in the next 50 years."

The old-line regional Bell operating companies (RBOCS), which until recently enjoyed steady revenue growth, are entering a period of increased market competition from wireless services, cable operators, and IP Telephony. For U.S. carriers, as a whole, their continued financial health will rely heavily on cost management. In face of declining service revenues, U.S. carriers will need to control both their capital and operational costs if they are to remain profitable. Carriers also need to develop strategies that will help them to con-tinue to grow their data services, thereby offsetting losses in voice. Even with strong data growth, it remains to be seen whether service revenue levels will ever return to those of 2001, reports In-Stat/MDR.

As the number of plain old telephone service (POTS) lines dwindle, digital subscriber line (DSL) technology will become increasingly important for U.S. carriers. DSL remains the primary method of broadband service for U.S. car-riers. According to In-Stat/MDR research, SBC and Verizon account for over half of all DSL lines in use.

For U.S. carriers, business data services will be a major growth area, as busi-ness needs for these services are continuing to expand. Data services include both the old private line standard and newer Internet access services. The outlook for traditional voice services is bleak, however. According to In-

Stat/MDR, the total long-distance voice market is on a downward spiral, with double-digit decreases in 2003 and 2004. The local market, although overall not as competitive as long distance, will see revenue declines of 4 to 6 percent over the next several years.

1.2   THE BROADBAND WORLD

As wirelines decline, the need for secure, speedy, and on-demand video, voice, and data services is rising. This "triple play" is propelling cable multiple service operators (MSOs), traditional phone carriers, and the consumer electronics industry to develop and distribute the means to transmit information to users worldwide.

By 2008, over 15 percent of households worldwide will have some type of high-speed broadband connection, predicts ABI, a technology research firm located in Oyster Bay, New York. The highest share of households will be in North America, followed by Western Europe, Asia-Pacific, and the rest of the world. ABI also finds that while cable broadband is leading in the United States, the worldwide DSL market share as of 2002 is around 60 percent, whereas cable broadband holds about 40 percent of the market.

But the desire to acquire more subscribers, while retaining existing ones, will spearhead the bundling and inter-reliance of "boxes" with any one or more of the triple play services in more creative ways over the next few years, leading to attractive price points and inventive services.

In the past, cable MSOs and consumer electronics vendors have had a disconnect in the way that they have offered services and products to consumers. "There always existed the 'chicken or the egg' arguments as to whether consumers should buy the products first based on the product's standalone features, or should the products be developed first and be made available for subscription to one or more of the triple play services," says Vamsi Sistla, ABI's director of broadband research. "Now, the unlikely bedfellows are seeing one another as necessary for survival."

Although worldwide digital cable households made up less than 9 percent of cable households in 2002, this share will grow continuously to reach just over 20 percent by 2008, forecasts ABI. However, this figure will represent only 7 percent of all the worldwide households as of 2008. The digital broadcast satellite (DBS) share of worldwide households will be over 12 percent in the year 2008.

Video-over-DSL will be the new kid on the block, with U.S. incumbent local exchange carriers (ILECs) and competitive local exchange carriers (CLECs) charging ahead with aggressive deployments to fend off cable's triple play offering. Even with higher growth rates, North American household video-over-DSL penetration rates will be trailing those of the Asia-Pacific region by 2 million, in the year 2008, forecasts ABI.

1.2.1 Broadband Over Power Lines

Although cable and DSL connections are the current broadband access leaders, a newer technology could prove attractive to millions of potential users, particularly residential and small business customers. If it ever reaches market, broadband over power lines (BPL) would turn every home and office electrical outlet into an always on Web connection, potentially providing stiff competition to cable modem and DSL service providers. The technology "could simply blow the doors off the provision of broadband," FCC chairman Michael Powell stated earlier this year.

BPL works by injecting data into medium-voltage power lines. Amplifiers are required at intervals along each line to keep signal strength at an acceptable level. Conventional fiber optic or copper phone lines are used to bypass high-voltage lines, which are too electrically disruptive to carry data. The carriers believe that ubiquitous BPL would provide broadband service to customers, including rural homes and businesses not currently served by cable modem or DSL providers, at comparable data speeds.

BPL is a viable technology, says Alan Shark, president of the Power Line Communications Association, a trade group located in Arlington, Virginia. He notes that BPL's technical hurdles, such as passing signals through transformers, have been largely overcome. The companies are now focusing on BPL's business case. "They're trying to figure out how to bring [BPL] to the home in the most cost-effective manner."

Despite its potential, BPL faces opposition in the wireless community. BPL's strongest opponent is the American Radio Relay League (ARRL), the national association of amateur radio operators. If widely deployed, BPL would represent "spectrum pollution" on a level that's "difficult to imagine," says Jim Haynie president of the 163,000-member ARRL, which is headquartered in Newington, Connecticut.

Haynie maintains that data signals radiated by power lines will seriously degrade amateur HF and low-VHF communications, both data and voice, at frequencies ranging between 2 and 80 MHz. He notes that BPL interference could also seriously affect national homeland security efforts. Many military, police, and public service radio users operate in the same 2 to 80 MHz spectrum range, and some of these organizations are presently unaware of BPL's potential threat. "In terms of interference potential on HF and low-VHF frequencies, nothing is on the same scale as BPL," says Haynie.

Haynie notes that BPL technology already has been deployed in some European countries and that hamoperators there have experienced interference from the systems. He adds that Japan—responding in part to concerns expressed by its amateur radio community—decided last year not to adopt the technology because of its interference potential. Shark, however, maintains that the ARRL's fears are overblown. "In the tests so far, there has been no interference," he says. However, Shark does admit that the "potential perhaps exists that if somebody had a [ham radio] rig right by a transformer, there could be a potential [for interference]."

Haynie, however, dismisses Shark's claim. He notes that recent field tests, conducted by the ARRL's lab manager in BPL trial communities in Maryland, Virginia, Pennsylvania, and New York, showed strong and sustained interference across all affected ham bands. "Anyone seeing these BPL signals for megahertz after megahertz for miles along a power line should be convinced that BPL—even operating at the present FCC limits—poses a serious threat to all HF and low-VHF communications."

1.3   THE UPCOMING MOBILE STALL

Given the fact that POTS is in decline, it only makes sense that the mobile phone service market must be soaring. And it is, although even runaway growth has its limits.

The next five years will see a dramatic slowing of worldwide cellular subscriber growth, reports In-Stat/MDR. However, despite much noise about a pending catastrophic slowdown in subscriber growth, there will be more than 931 million new subscribers over the next 5 years. By 2007, the total worldwide wireless population will exceed two billion subscribers.

A recent report issued by In-Stat/MDR finds that, although China continues to lead the world in overall subscriber growth, the new percentage growth leaders can be found in Southern Asia and Southeast Asia. "It is rather remarkable that the fastest numerically growing country, China, is trailing Africa, Eastern Europe and the Middle East in compound annual growth rate," says Ken Hyers, a senior analyst with In-Stat/MDR. "However, the fact that Western Europe and Scandinavia bring up the rear, as they did in previous 2002 to 2006 forecasts, should be no surprise." Indeed, In-Stat/MDR finds that Western Europe's growth virtually stops during the 2002–2007 forecast period, with a compounded annual growth rate of 1.2 percent. This statistic is hardly surprising in light of the fact that mobile phone penetration rate in 2007 will be 83.6 percent.

Meanwhile, research from Yankee Group, a Boston-based technology research company, shows that the U.S. wireless industry is approaching maturity with impressive speed. Only three years ago, the U.S. wireless industry was one of the few remaining emerging high-growth sectors. However, with almost 50 percent penetration, North Americans now treat wireless like a utility rather than a novelty.

"The U.S. wireless industry is facing the threat of becoming like the airline industry with high fixed costs, low variable costs, a perishable product and cutthroat competition," says Roger Entner, Yankee Group's wireless/mobile services program manager. "These conditions make it easy for industry participants to behave in a way that has potentially disastrous consequences in the long run. Airline travel is cheaper than ever before, but few customers are happy with the experience."

Given the present state of wireless competition, it's only a matter of time before unlimited calling plans are available nationwide. Wireless number portability (WNP), which began in 2003, likely will be a catalyst for this trend, according to the research firm.

1.4 FOURTH-GENERATION MOBILE SERVICES

With the wireless industry looking for new ways of boosting both subscriber numbers and usage, most carriers are already planning fourth-generation (4G) networks. As a result, mobile phone networks are destined to become much faster and more uniform over the next several years. Today's networks, restricted by low bandwidth and a patchwork of incompatible standards, will give way to an interoperable system that supports an array of devices and offers seamless roaming. Imagine a wireless world in which networks provide broadband data and voice, giving users high-quality audio, Internet, and even video services. Users can go anywhere in the world and automatically be handed off to whatever wireless service is available, including cellular, satellite, and in-house phone systems.

Higher-speed third-generation (3G) mobile services has now rolled out, but 4G technology is waiting in the wings. Upcoming 4G services, intended to provide mobile data at rates of 100 Mbits per second or faster, could begin arriving as soon as 2006. According to the Fourth-Generation Mobile Forum, an international technical body that's focusing on next generation broadband wireless mobile communications, the technology is about to undergo explosive growth. In 2000, only eight organizations were involved in 4G research and development. In 2002, over 200 companies and research institutions were conducting 4G projects. By 2008, over $400 billion will be invested in 4G services.

Several major carriers have already started 4G testing. Japan's NTT DoCoMo, for example, has been conducting research on 4G mobile communications technology since 1998. In indoor experiments conducted in 2002, NTT DoCoMo's 4G system demonstrated maximum information bit rates of 100 Mbps for the downlink and 20 Mbps for the uplink.

Emerging 4G technology promises to converge wireless access, wireless mobile, wireless local area network (WLAN), and packet-division-multiplexed (PDM) networks. With PDM technology, for example, a single integrated terminal using a single global personal number can freely access any wireless air interface. Additionally, PDM radio transmission modules are fully software definable, reconfigurable, and programmable.

NTT DoCoMo is currently conducting research into a technology known as variable spreading factor-orthogonal frequency and code division multiplexing (SF-OFCDM), which has the power to transmit at speeds of up to

100 Mbps outdoors and up to 1 Gbps indoors. Basic functionality has already been verified for this technology and NTT DoCoMo is now involved in actual field experiments. The firm is also building a mobile IP network specifically for packet data that supports seamless service between the company's mobile service and a variety of other networks (such as WLANs) to provide an enhanced online experience with reduced network cost.

NTT DoCoMo is also working on an entirely new system concept that will have the power to do away with base stations entirely by allowing terminals to interconnect directly. This company is also investigating versatile mobile networks where base stations will have the ability to install themselves automatically to achieve a network that actually thinks for itself.

1.5 MODULAR COMPONENTS

As mobile operators race to provide ever more sophisticated and complex services, companies must soon redraw their IT architectures and adopt modular software components in order to market new services quickly and cheaply.

For example, many mobile operators find it hard to market their products quickly because of a complex and inflexible IT architecture that forces them to develop many parts of each new product almost from scratch. Product developers who can't reuse components across applications must constantly reinvent the wheel, asserts a study by McKinsey & Company, a management consulting firm based in New York.

Component reusability remains rare because speedy growth ruled the telecom industry during the boom years of the late 1990s, when companies had neither the time nor the inclination to consider which software components could be reused in other products. The quickest way to get out new offerings was to patch the existing architecture by forging connections between whatever systems immediately needed them. The result was an increasingly complex, spaghetti-like architecture littered with incompatible stand-alone systems. Such systems were based on software from a number of vendors and often using a variety of incompatible data formats, such as customer databases with different sets of vital statistics.

To illustrate the problem, the McKinsey report offers the example of a mobile device restaurant finder. An operator developing such a product starts by defining its characteristics, how to deliver the information to the subscriber (such as SMS, the Multimedia Messaging Service or the mobile Internet), and a pricing scheme. The programmers then work on creating the applications, databases, and interfaces. All of this makes for an arduous process, involving thousands of hours of coding and adding greatly to the project's cost.

Such a product also requires a variety of support features, including a restaurant database, customer profiles, and systems for locating and billing subscribers. Unfortunately, such features aren't always readily available. Information about customers, for example, will almost certainly be spread over multiple databases and applications. Programmers may be able to access it, but they will need time to understand the code and data structures of legacy applications, as well as time to create interfaces to legacy databases and to combine and match customer information from many different sources. As a result, project's programmers will not focus on creating a differentiating customer experience but simply will focus on getting the basics right. "To begin constructing those support functions, mobile telecom companies should reorganize their information systems into reusable building blocks, or components," notes the McKinsey study. "Assembling and reassembling them into the basic elements of a mobile product then becomes a lot less time-consuming and costly."

There are multiple benefits to a component-oriented IT architecture, notes the McKinsey study. An architecture with reusable components would permit a team developing a mobile product to scroll through a company's database of services and to pick what it needed straight off the shelf or to tweak existing elements of the service. The team would then be free to concentrate on developing the product's features. "This approach, we believe, will become common in mobile telecommunications over the next few years," notes the study. "Judging by the results achieved in other industries, mobile operators could reduce the time to market of a new product by 30 percent and cut the cost of integrating it into an existing system by 60 to 70 percent."

1.6   A CONSIDERATE TELEPHONE

Besides allowing people to communicate in entirely new ways, emerging technology is also enabling individuals to interact with phone services in new and innovative ways. Telephones today, whether landline or wireless, are our cruel masters. They command our attention and don't care if we are eating dinner, engaged in a crucial business meeting, or watching a movie. But people may soon gain some control over their phones, thanks to pair of Carnegie Mellon University researchers who are working on a phone that could learn when—and when not—to summon its user.

The technology, which is being developed by researchers James Fogarty and Scott Hudson, utilizes tiny microphones, cameras, and touch sensors to monitor a phone user's activity level and body language. Software is used to monitor the various input devices and to determine whether the individual is too busy to bother with an incoming voice call or text message. "The idea is to get the telephone to act more as an assistant than a tool," says Fogarty, a Carnegie Mellon doctoral student.

Under one approach, the sensing infrastructure would be independent of the communications device but able to communicate with the unit. "You could instrument an office, for example," notes Fogarty. Homes, cars, and other locations could also be equipped with sensors to monitor their inhabitants' activities. An alternative approach would be to build the infrastructure, perhaps including a camera, microphone, and movement-detecting accelerometer, into the communications device itself. "That would probably be the least expensive approach," says Fogarty.

The researchers recently tested their technology, using 24 sensors, on four individuals engaged in work activities. The subjects were asked at random intervals, on a five-level scale ranging from "highly interruptible" to "highly not-interruptible"—how willing they were to be bothered with a phone call. The researchers then correlated the subjects' preferences with their behaviors.

Not surprisingly, the test showed that the subjects were least likely to appreciate an interruption while typing on a keyboard, talking on a phone, or speaking with someone else in the office. What did surprise Fogarty and Hudson, however, was the fact that the computer was slightly more accurate than human observers at predicting when an individual was willing to be interrupted. The computer successfully predicted the subject's preference 82 percent of time while humans managed only 77 percent accuracy.

Hudson believes that the test results will carry over successfully to the real world, particularly in business settings. "I'm confident that the results will apply to communications-oriented people such as managers," says Hudson a professor at the Carnegie Mellon Human-Computer Interaction Institute. Hudson admits, however, that he's not yet sure how well the technology will work with people in other types of jobs, as well as consumers. "I suspect that something slightly different will be needed, particularly for task-oriented workers," he says.

Fogarty and Hudson believe their technology could provide benefits beyond basic voice call convenience. The system could, for example, prioritize incoming instant messages based on the user's current activity—sending critical messages through immediately, delaying others to a more convenient time, and jettisoning spam. "Quite simple sensors will do the job," says Hudson.

The researchers first plan to integrate the technology into a computer-based instant messaging system. If that implementation proves successful, they will next target landline and wireless phones. "There's no technological roadblock that would prevent this technology from being deployed within a couple of years," says Hudson.

The researchers' ultimate goal is really quite simple, says Hudson. "We're out to prove that mother was right: it's not polite to interrupt while someone else is talking."

1.7 E-MAIL LEADS TO INSTANT MESSAGING

E-mail is a dominant data communications service, although its future is becoming less certain, due to rising spam and instant messaging usage. With more than 500 million business E-mail users worldwide today and over 20 billion spam messages expected to be sent daily worldwide by 2006, according to IDC statistics, the impact on business communications is huge.

IDC estimates that spam represents 32 percent of all external and internal e-mail sent on an average day in North America in 2003, up from 24 percent in 2002. The rising torrents of spam are reducing e-mail's usefulness by forcing users and IT staff to expend additional time and energy to identify, delete, and prevent spam from clogging in boxes. "To keep e-mail at the collaboration center stage, e-mail proponents will need to do a better job of helping end-users manage e-mail and use other collaborative tools in conjunction with e-mail," says Mark Levitt, research vice president for collaborative computing at IDC.

The value of instant messaging's immediacy and presence awareness is being noticed more widely in the workplace. However, instant messaging is becoming more similar to e-mail in terms of corporate requirements for tracking and archiving of messages.

Long favored by gossipy teenagers, instant messaging is now donning a suit and showing up for work. The software, popularized by programs such as AOL Instant Messenger, Microsoft's MSN Messenger, Yahoo Messenger, ICQ, and IRC, is being adopted—albeit often reluctantly—by a rapidly growing number of enterprises. "I think you're going to see IM use grow much faster than e-mail use," says Michael Osterman, president of Osterman Research, a technology research company in Black Diamond, Washington.

ComScore Networks, a Reston, Virginia-based audience ratings company, estimates that the number of work-based instant messaging users rose 10 percent during the first six months of 2002, reaching 17.4 million active users. "The same services that people have early-on adopted for use at home, mainly for social reasons, are now catching on at work," says Max Kalehoff, a senior manager at ComScore.

Unlike e-mail, instant messaging can deliver messages directly to a recipient's desktop, where it's likely to receive immediate attention. The technology can also be used for customer support and to simultaneously send messages to dozens or even thousands of users. With the arrival of instant messaging software into the business mainstream, many CIOs are concerned that instant messaging will open yet another door through which hackers can crawl. They also worry that instant messaging will sap productivity.

Despite the questions, instant messaging's popularity has drawn a variety of vendors into the field, with easily downloadable tools often appearing at enterprises that have yet to adopt a formal instant messaging strategy. "Most of the IMing at work is done through the big-brand instant messaging services," says Kalehoff. Osterman notes that enterprise adoption of instant messaging technology is lagging far behind employee demand. "Only about 30 percent of companies have established a corporate standard for IM," Osterman says. On the other hand, he notes, about 85 percent of companies have some level of instant messaging activity.

Capitalizing on the fact that their products already contain an instant messaging-type technology, conferencing and collaboration software vendors such as Groove Networks and Lotus Software are also entering the field. Lotus, which sells Sametime collaboration software, has already gained a solid foothold in the enterprise-grade instant messaging market. "Among organizations that have actually established a standard, about 60 percent have established Sametime as the standard," says Osterman. Unlike consumer instant messaging software, Sametime provides several enterprise-class management and security features, such as integration with corporate directories and encryption. Also entering the field are numerous pure-play instant messaging startups, such as Bantu, Ikimbo, and Jabber. These companies hope to beat the competition with instant messaging multimedia messaging tools that span multiple platforms.

The thought of employees flinging unencrypted messages through public networks, however, is enough to give almost any CIO the willies. The idea that external instant messaging senders may be able to toss viruses and other types of destructive code into an enterprise is at least equally chilling. Although most enterprise-grade instant messaging softwares offer some type of security mechanism, primarily encryption, many consumer-grade products—the kind brought in by employees without the IT department's knowledge—don't. "One of the problems with traditional consumer-grade clients is that they can open a hole in the firewall," says Osterman. "Then you have a path for viruses and malicious codes."

Compatibility problems also plague instant messaging; a universal standard is still somewhere in the future. Presently, most instant messaging products can't display messages from competing systems. "[AOL IM] is a popular service in some departments, and MSN is popular in other departments," says Kalehoff. "The problem is they don't talk to one another."

In addition to the security and compatibility traps of instant messaging, CIOs must also worry that the technology will eat into productivity. After all, repeatedly pausing to answer messages and swat nuisance IM pop-ups isn't a great way to focus. "You can specify that you're busy, but you have an extra step not to be disturbed," says Osterman. Employees are also likely to use the technology to chat with family and friends. "This is something that could be used as a time waster," says Osterman.

Many businesses also deal with IM's legal implications. This is particularly true in the financial industry, where Securities and Exchange Commission regulations require securities companies to record and log both instant messages and e-mails. Although most enterprise-grade instant messaging products, such as Sametime, provide archiving capabilities, many financial industry CIOs would simply prefer to skip the complex job of tracking individual instant messaging pop-ups. These CIOs have either banished instant messaging from their organization or limited its use to purely administrative functions. "We're not really communicating dollar figures or anything like that," says Robert Stabile, senior technology officer at investment company J.P. Morgan Partners in New York City.

In fact, given the strong likelihood of technical and management headaches, CIOs at all sorts of organizations would simply like to exile instant messaging technology. Many already have. According to Osterman Research, 22 percent of companies block IM traffic from their network.

APL, a 12,000-employee containerized shipping company based in Oakland, California, put the hammer down on instant messaging when employees began installing consumer-grade client software on their desktop. "We started to see that it was eating up bandwidth; we started seeing file transfers via instant messaging," says Van Nguyen, APL's IT security director. After determining that instant messaging was more of a convenience tool than an essential business application, Nguyen and senior managers pulled the plug. "We have implemented a corporate-wide security policy to disallow instant messaging clients—period," he says.

Although banning client software is an easy way of dealing with instant messaging's problems, this move may also be shortsighted. Many enterprises that have adopted instant messaging are beginning to appreciate the technology's potential to actually boost productivity. Adopting a formal instant messaging strategy also lowers the likelihood that employees will sneak in less secure consumer-grade products.

When the employees at Avnet Computer Marketing want to send an important message to colleagues or customers, they don't necessarily reach for a phone or e-mail. More often than not, the information is typed into an instant messaging application. "You can just bounce a couple of lines across to somebody and get an answer," says Dave Stuttard, vice president of application solutions for the Tempe, Arizona-based computer products distributor.

At Avnet Computer Marketing, about 500 employees use instant messaging for a variety of tasks. In one pilot project, for example, customers can use instant messaging to contact technical people at the company. The software also reduces the need to place costly international phone calls. It's too early to tell just how much money instant messaging is saving, Stuttard says, but he's sure that the technology is having a positive effect on the bottom line. Stuttard says that, when all is said and done, the company hopes to reduce its number of voice mails and e-mails, while providing faster turnaround on decisions.

Instant messaging's cost savings potential hinges mostly on how the technology is used. "If it was used primarily as a replacement for long-distance calls," says Osterman, "then the savings in telephone charges could be substantial in a large organization." Similarly, if the technology serves as an e-mail replacement or supplement, "there could be some savings in disk storage and related requirements," he says.

As time goes by, even Nguyen is contemplating a return to instant messaging—but only under tightly controlled conditions. "We're looking to internal instant messaging servers," he says. APL's planned approach would place instant messaging activities into an encrypted, VPN-type environment that would encompass only employees and selected external parties. "If it's a business requirement, definitely we would allow external partners to communicate with us," says Nguyen.

As instant messaging becomes a deeply ingrained technology, messaging functions are likely to begin popping up inside all sorts of business-oriented applications, ranging from word processors to accounting applications. "For example, you might see a future version of Microsoft Office that contains instant messaging functionality," says Osterman. (Houston-based Advanced Reality already offers tools for adding collaboration to any application.)

One possible Microsoft strategy would be to add instant messaging support to .Net Server, its latest server operating system. Code-named Greenwich, Microsoft's instant messaging software will provide a variety of multimedia tools to connect users in real-time. "Greenwich envisions building on core presence capabilities to deliver instant messaging, voice, video and data collaboration as a standards-based, extensible real-time communications solution," says Bob O'Brien, group product manager of Microsoft Windows .Net division. In the meantime, Yahoo has announced the release of its corporate instant messenger, which will include the capability to integrate with corporate directories and some applications.

Increasing enterprise adoption of instant messaging is also likely to lead to new uses for the technology. Avnet's system, for example, allows technicians to communicate with customers on particularly difficult problems. NEC Solutions' Visual Systems Division, an Itasca, Illinois-based display products vendor, is using instant messaging software to directly assist customers. "They can instant message their customer support rep and get the information they need instantaneously," says Fran Horner, director of the division's service sales group. The company's instant messaging system even has the ability to transmit diagnostic software and fixes directly to a user's desktop.

Ultimately, enterprise instant messaging will span an array of platforms, allowing users to conveniently contact people anytime, anywhere: on a desktop PC, personal disital assistant (PDA), mobile phone, or other connected device. Several vendors, including Bantu and Jabber, already provide software with a multiplatform capability.

1.8 FUN AND GAMES

Telecom services don't only carry voice and information. In the new Telecosmos, entertainment is an important diversion for telecom users and a profit center for service and content providers.

Wireless gaming is well on its way to becoming a mass-market phenomenon. Wireless games currently top the list of applications downloaded to cellular phones. IDC, a technology research firm located in Framingham, Massachusetts, expects the number of wireless gamers to grow from 7.9 percent of all U.S. wireless subscribers in 2003 to 34.7 percent, or 65.2 million users, by 2008.

"In 2003, U.S. wireless carriers cleared a major hurdle in delivering wireless games to subscribers, demonstrating that wireless gaming is a viable business," says Dana Thorat, a senior research analyst in IDC's wireless and mobile communications service. "Carriers plan to aggressively promote wireless games to their subscribers while offering new line-ups of compelling titles, including those that support multiplayer and limited 3-rendering."

So far, carriers have pursued mass-market strategies in targeting games to a broad spectrum of consumer demographics. The key to wireless game success has been mostly related to strong brand and game title recognition. Popular wireless games in 2003 have included Jamdat's *Jamdat Bowling*, Activision's *Tony Hawk's Pro Skater*, Eidos' *Lara Croft Tomb Raider*, and Gamelofts' *Tom Clancy's Splinter Cell*.

For the carriers, getting to market quickly with compelling game titles is the key to unlocking the new revenue opportunities of wireless gaming. The growth of this market will not only depend on the infusion of download-capable handsets but also more effective merchandizing, such as recommendation engines, opt-in e-mail, and five-digit short code marketing, as well as various upselling and cross-selling techniques using other mediums such as banner ads on online game sites.

1.9 FLYING PHONE SERVICE

Talking on a mobile phone while flying on a commercial airliner could soon become reality, at least if one company has its way. AirCell has developed a technology that would allow mobile phone users to place and receive calls as if they were still on the ground. The system uses airliner-mounted radio transceivers to connect callers with any of 135 antenna sites across the U.S. The company was awarded a U.S. patent for its technology last month.

Compared with earlier technologies, which involved placing the equivalent of a full terrestrial mobile phone base station aboard an aircraft, AirCell's approach reduces the size, weight, and cost of equipment required to provide phone service on board an aircraft. "The patented concepts allow all the functionality provided to the cellular user in-flight to be controlled by the network rather than the airborne station, simplifying the addition of features and migration to future cellular technologies, and it also provides a novel way to manage the radio frequency environment in the aircraft to prevent interference," says Ken Jochim, vice president of engineering and operations for the Louisville, Colorado-based company.

The company notes that rigorous testing will be required to satisfy concerns relating to possible interference with the aircraft's communication and navigation systems. AirCell says it has entered into discussions with the FCC and FAA to ensure all requirements are properly met.

Although airlines are reluctant to admit the fact, an unknown percentage of passengers actually do use their mobile phones—albeit stealthily—while flying. "Many passengers use their Blackberry devices on planes as well," says Edward Rerisi, an analyst at research firm ABI. "These technologies work, but coverage is often spotty and the exact effects on the aircraft's communications are still unproven."

Flying phone users can also wreck havoc on terrestrial base stations. "When airborne, a single mobile handset may be able to transmit to multiple base stations," says Rerisi. "This poses a particularly challenging problem with CDMA networks, but reportedly less so with TDMA, GSM and analog networks."

Although AirCell's technology addresses the technical challenges posed by airborne mobile phone use, the company has yet to unveil the system's business model. "Billing will be a challenge," predicts Rerisi. "What about pricing plans? Surely the airlines will want to share in the revenues garnered during in-flight calls."

AirCell states that it is working toward getting its technology certified for commercial air transport aircraft and that discussions are in process with several airlines to finalize plans for a trial program. Rerisi describes AirCell's system as a "plausible" technology, although "regulatory and business challenges may prevent the technology from leaving the ground."

1.10 SPEECH INTEGRATION

Speech integration is the technology that adds voice services to enterprise phone systems and Web sites. The speech recognition market has suffered along with the entire telecommunications industry during the past few years, but the market now appears to be poised for renewed growth. "There are several signs that the speech recognition industry is maturing," says Steve Cramoysan, a principal analyst with Gartner, a technology research firm located in Stamford, Connecticut. Many implementations provide proof that solutions that use speech recognition can deliver business value, as cost savings or improved customer service."

Speech recognition performance has improved versus the products available only a couple of years ago. As a result, it's becoming increasingly difficult for vendors to differentiate their wares purely on the basis of speech recognition success rates. Internet-based applications and standards, such as VoiceXML, are gaining market share, providing an increasingly distributed architecture that allows companies to leverage their investment in speech technology and allows services providers to offer speech recognition services to enterprises. "The clear market leaders today are Nuance and ScanSoft. Entry by Microsoft, IBM, and Intel into the market is providing significant momentum, and further changes in the vendor landscape are to be expected," notes Cramoysan.

Perhaps the most important use of speech recognition technology over the next several years will be in speech integration systems that automate and streamline enterprise phone systems. Speech integration technology is nothing new, as any telephone caller who has ever barked back responses to a seemingly endless series of voice prompts can testify. But an improved generation of speech integration software, based on more powerful processors and emerging Internet-focused standards, promises to make the technology more useful and cost effective.

Until recently, organizations tended to shy away from speech integration because of the technology's complexity and cost. "I had one client who had 60 people on its [speech integration] project," says Elizabeth Ussher, an analyst who covers speech technologies for Meta Group, a technology research firm located in Stamford, Connecticut. Today, preconfigured speech templates, drop-in objects, and other packaged tools make speech integration development less burdensome. Hardware improvements, particularly speedier processors, also help make speech integration a more practical technology. "Speech recognition is now very widely deployable," says Ussher. "I'm seeing clients with a return on their investment within three to six months."

Yet another reason for increased interest in enterprise speech integration can be found in the almost exponential proliferation of mobile phones, PDAs, and other portable wireless devices. Speech input/output is an attractive alternative to cramped keyboards and miniscule displays. "If I'm on my mobile phone while driving my car, I'm not going to push buttons for my account number," says Ussher. "I'm going to wait for an agent—living or virtual."

Dollar Thrifty has been using speech integration to handle some of the more than 1 million calls it receives each year from "rate shoppers"—bargain hunters who phone several different car rental companies in search of the best deal. "Many of the folks who call are just interested in checking rates," says Bob Dupont, vice president of reservations for Thrifty. "They aren't interested in making a reservation; they just want to get information for comparison purposes."

To free its call center staff from the burden of handling routine data lookups, Dollar Thrifty installed SpeechWorks International's software at its Thrifty division. The system lets callers check rental rates and availability at airport locations by talking with a virtual call center agent. "It's a very natural, realistic interchange," says DuPont. The software also automatically adapts to unique requirements, such as providing personalized rates for members of Thrifty's loyalty program.

After checking rates and availability, callers who decide to make a reservation are seamlessly transferred to a live agent. A screen "pop" automatically appears on the agent's display, presenting all the information the caller provided during the speech interface dialogue. DuPont estimates that 35 percent of calls to the company's toll-free number go through the speech integration system. In addition, speech integration has not hurt Thrifty's conversion rate— the number of people calling for a quote who ultimately make a reservation.

Deploying the system wasn't especially difficult, he adds. "Just the normal tweaking of the application and getting the voice recognizer to work better. Once we got through the first 90 to 120 days, it became apparent that we had a very solid application." Uptime has been more than 99 percent, which is a critical factor, says DuPont. "If it were to go down, we certainly would be understaffed."

Enterprises looking into speech integration face two basic technology choices. The oldest and simplest type of speech integration—"directed dialogue" products—prompts callers with a series of questions and recognizes only a limited number of responses, such as "yes" and "no," specific names, and numbers. A new and more sophisticated approach—"natural language"—to speech integration handles complete sentences and aims to engage callers in lifelike banter with a virtual call center agent. The technology is also more forgiving of word usage. "If a customer calls Thrifty and asks about rates from JFK Airport in New York, they might say 'JFK' or 'John F. Kennedy' or 'Kennedy Airport,'" says SpeechWorks cofounder and CTO Michael Phillips. "The system has to be prepared for the different variations that might be used."

Directed dialogue tools, although less expensive than natural language systems, suffer from their limited recognition capabilities. As a result, they are mostly used for simple applications, such as automated switchboard attendants or credit card activators. Natural language systems, such as the type used by Dollar Thrifty, have a wide range of applications, including product and service ordering, telebanking, and travel reservation booking.

A pair of emerging technologies—VoiceXML and Speech Application Language Tags (SALT)—are also helping to advance voice integration. Both rely on Web technology to make it easier to develop and deploy speech integration applications. VoiceXML is an XML extension for creating telephone-based, speech-user interfaces. VoiceXML lets developers create directed dialogue speech systems that recognize specific words and phrases, such as names and numbers. That style of interface is well suited to callers who have no screen from which to select options. SALT, on the other hand, provides extensions to commonly used Web-based markup languages, principally HTML and XHTML. It makes such applications accessible from GUI-based devices, including PCs and PDAs. A user, for example, might click on an icon and say, "Show me the flights from San Francisco to Boston after 7 p.m. on Saturday," and the browser will display the flights. Both specifications aim to help developers create speech interfaces using familiar techniques. "You don't have to reinvent the wheel and program a new interface to get speech recognition access to your data," says Brian Strachman, a speech recognition analyst at technology research company In-Stat/MDR.

Although most people think of speech integration in terms of customer self-service, the technology can also be used internally to connect an enterprise's employees and business partners to critical information. Aircraft mechanics, for example, can use speech integration to call up technical data onto a PDA or notebook screen. Likewise, inventory takers can enter data directly into databases via speech-enabled PDAs, without ever using their hands. The Bank of New York, for example, has tied speech recognition into its phone directory and human resources systems. Using technology supplied by Phonetic Systems, the bank operates an automated voice attendant that lets callers connect to a specific employee simply by speaking that person's name. However, in the event of a major emergency that requires entire departments to move to a new location, the employees can call into the system to instantly create updated contact information. The information then becomes available to anyone calling the bank's attendant.

The speech-based approach is designed to help bank employees resume their work as soon as possible, even before they have access to computers. "The automated attendant was already connected to our back-end systems," says Jeffrey Kuhn, senior vice president of business continuity and planning. "We simply expanded the number of data fields that are shared between the Phonetic's product, our HR system and our phone directory system." The biggest challenge Kuhn faced in deploying the technology was getting it to mesh with the bank's older analog PBX systems. That problem was eventually solved, although the interface ports on the old PBX units must now be manually set, which is a minor inconvenience.

Speech integration's primary benefit for callers is convenience, since the technology eliminates the need to wait for a live agent. Problems handling foreign accents, minor speech impediments, and quirky word pronunciations have largely faded away because software developers have given their products the capability to recognize and match a wider array of voice types. "Every four to five years, speech technologies improve by a factor of two," says Kai-Fu Lee, vice president of Microsoft Speech Technologies. Dollar Thrifty's DuPont says his company's internal research has found an end user satisfaction level of around 93 percent. "It either met or exceeded their need to get information, and they had an improved perception of our company," he says.

For enterprises, speech integration's bottom-line benefits include cheaper user support and data access. DuPont says his system paid for itself in less than one year, lopping about 45 cents off the cost of each incoming call for Thrifty. Bank of New York's Kuhn estimates that his system handles the work of five full-time employees. Still, despite the potential benefits, enterprises shouldn't view speech integration as a panacea to their rising call center costs. The technology itself requires constant attention, which adds to its base cost and detracts from potential savings. "It's labor intensive," says Meta Group's Ussher. "It's not like a washing machine that runs on its own. It's a technology that requires constant tweaking, pushing and updating." DuPont warns potential users not to consider speech integration as solely an IT issue. Because the technology affects a wide range of business processes, he believes that it's vital to garner enterprise-wide support. "I would certainly recommend getting all the stakeholders involved," he says. "When we put our system together, we involved people from many disciplines, including IT, HR, finance and telecom, as well as the reservations group."

Although speech integration will certainly become more capable and self-sufficient in the years ahead, few observers believe the technology will ever fully replace living, breathing call center agents. In-Stat/MDR's Strachman says that speech integration will primarily be used to eliminate call center grunt work, such as the recitation of fares and schedules, and to give end users a new way to access critical data. The handling of complex issues, such as technical support, will probably always require access to a live expert. "For call center agents to stay employed, they're going to have to be more highly skilled and trained than they are now," says Strachman.

1.11 TELEMEDICINE

New telecom service, hardware, and software options are opening the door to advanced video and data monitoring capabilities. The health care industry is leading the way in using these technologies to address real world problems. For example, research shows that substituting interactive video sessions for up to half of a visiting nurse's in-home meetings with postsurgical or chronically ill patients can be a cost effective way to provide care.

"Video visits are not a complete substitute for in-home nursing care," says Kathryn Dansky, a Penn State University associate professor of health policy and administration. "You are always going to need home visits because patients benefit from the personal touch." Still, a recent study led by Dansky found that, over a typical 60 days of care, savings of $300 per patient could be achieved by substituting video visits for seven in-home visits and $700 per patient was saved if half of the visits were made via advanced communication technology. "As the number of nursing visits increase, you can substitute more and more video visits if the purpose is to monitor the patient's health status," says Dansky. "Substituting an equal number of video and home visits can produce a major difference in the cost of the care." The sources of savings include less travel time and travel costs, fewer travel accidents, less car theft, and the ability to see more patients in the same amount of time.

Skilled nursing care in the home requires a registered nurse to drive to the patient's residence, conduct examinations and assessments, provide patient care and education, and then drive to the next patient's house. The process is time consuming, dangerous at times for the nurse, and expensive. To see whether new technology could help both patients and nurses without incurring additional costs, the researchers initiated a 24-month evaluation of the use of telecom as a supplement to skilled nursing visits for people with diabetes. Called the TeleHomecare Project, the effort was a partnership of Penn State, American Telecare Inc., and the Visiting Nurses Association of Greater Philadelphia (VNAGP), a large, urban, home health agency.

A group of 171 diabetic patients discharged from the hospital and referred to the VNAGP participated in the study. Half of them were randomly assigned to receive a patient telecommunication station in their homes, while the remaining patients received traditional in-home nursing visits. The patient station included a computer and monitor equipped with two-way voice capability and a video camera. A blood pressure cuff and stethoscope were also attached to the computer.

Using the patient station, which works over ordinary phone lines, the patient could see and talk with the nurses. The system also allowed the nurses to see and hear the patients and to take temperature and blood pressure measurements, listen to heart and lung sounds, and discuss diet and blood sugar results. Patients who used the telecommunications system scored higher on positive outcomes of treatment, had fewer rehospitalizations, and had fewer visits to hospital emergency rooms.

Dansky notes that, in general, the patients liked working with the telecommunications equipment. The stations gave patients a sense of security because they could keep in touch with their nurse at all times. Some patients even prepared for the video visits by fixing their hair and dressing up. Far from frivolous, such interest in self-care is an important indicator of vitality and personal responsibility.

The nurses, too, responded favorably to the technology, although three generations of telehomecare machines were introduced and tested during the study period. Dansky notes that the nurses found ingenious ways to deal with equipment failures. For example, if a patient didn't respond, they would hold up a sign that says, "Nod your head if you can see and not hear me." The nurses also used laundry baskets to take the equipment into homes so that thieves wouldn't see what they were doing. There were no thefts during the project and no break-ins, even though some patients resided in crime-ridden areas of the city.

Dansky sees many possibilities for broader application of the telecommunications systems. She is currently working with Sun HomeHealth to study whether the systems can aid nurses in helping patients manage their medications, especially when there is a danger of drug interactions. She also sees the possibility of physical therapists using the system to supervise family members or aides helping patients exercise in their homes. Dieticians could also use the system to supervise meal planning and preparation.

1.11.1 Health Monitoring

Sophisticated health monitoring services are beginning to allow patients to receive quality health care even while at home. Companies like Philips Medical Systems are pioneering this field. Philips and HomMed have one objective, although their approaches to in-home health monitoring are different. HomMed's system is fully wireless, whereas Philips offers a combination wireless-and-wireline configuration.

With HomMed's system, a portable instrument console sits in a patient's bedroom or similarly convenient place. The unit accepts plug-in measurement devices that collect patient health data, such as heart rate, blood glucose levels, and body temperature. The information is sent via the SkyTel pager network to a central station located at a hospital or a clinic where a clinician reviews it. A wireline connection is available as a backup.

A key reason that HomMed is using wireless technology is that one of every eight patients doesn't have a home phone, according to HomMed CEO Herschel Peddicord. He also cites the technology's portability as a plus. "HomMed's system can be taken on the road and used in a hotel room or a vacation home, which is particularly handy for elderly patients who travel to winter homes," he says.

HomMed is confident that wireless is the best way to move patient data from homes to clinicians; however, Philips says the combination wireless-and-wireline approach of its heart-care system is less expensive and more reliable and secure. "Phone lines are highly dependable, service is universal, and we don't have to reassure people that their data won't be intercepted by electronic eavesdroppers," says Steve DeCoste, business manager for Philips Medical Systems.

Critically ill heart patients usually don't travel much, DeCoste notes, so there isn't a great demand for portability. "We're not after that diabetic who's 25 years old and still rides his mountain bike," he says. Philips' system, which the company acquired last summer from Agilent Technologies Inc., uses battery-operated wireless measurement devices, such as weight scales and blood-pressure cuffs, that can be placed anywhere in a patient's home. "Since our patients tend to be very sick, it's important for the devices to be easily accessible at a bedside or other convenient location," DeCoste says.

The wireless devices allow patients to gather measurements daily—such as weight, blood pressure, pulse, and heart rhythm—that are vital to the ongoing management of congestive heart failure. The data are transmitted to a hub box, which collects the information and automatically sends it via a phone line to a clinician for examination.

1.11.2 Small Clinics/Hospitals

Monitoring patients after they have left the hospital is a vital part of follow-up care. Yet many small clinics and hospitals find it difficult to provide adequate outpatient support, primarily due to distance and budgetary limitations.

A new software package aims to give small institutions the technical means to improve outpatient follow-up by accessing servers located at larger hospitals. Using the software, small clinics and hospitals can access and use the data held by larger institutions to better track patients' medical and nutritional care and to set up automated prescription services.

Sponsored by EUREKA, the pan-European research and development consortium, the software is focused on nutritional follow-up. The technology can be used to monitor children with diabetes or other illnesses that require careful diet monitoring. For children with diabetes, the software can help minimize long-term complications, such as damage to the eyes, nerves, and kidneys.

"Nutrition is of critical importance for certain patients," says Bernard d'Oriano, managing director of Fichier Selection Informatique, the French company that's leading the project. Careful nutrition monitoring is also crucial because of Europe's rapidly aging population. "Elderly people can become malnourished very quickly, even if they are still eating, and can become critically ill within the space of three weeks," notes d'Oriano. Using the new software, a doctor could alert the hospital whenever they suspect a patient is in danger, and the hospital could help monitor the person's diet.

A companion software package allows doctors to send prescription orders electronically. "The doctor at the patient's bedside enters the prescription onto a laptop or a PDA and sends it via the Internet directly to the hospital's in-house pharmacy," explains Philippe Corteil, managing director of the venture's Belgian partner, Medical Business Channel. The pharmacy can then instantly dispense the medicine and keep an accurate account of both what is going to the patient and the stock remaining in the pharmacy. Simultaneously, the software enables the prescribing doctor to see what other pharmaceuticals have been prescribed for the patient and to be alerted if there is a potential medicine conflict.

For d'Oriano, joining the EUREKA project was advantageous in several ways. "Above all, it brought me the means—a loan as well as a Belgian technical partner with whom we were able to work—and a certain reputation and recognition," he says.

1.11.3 Monitoring on the Road

In California's Santa Cruz County, all of the ambulances have been equipped with cardiac monitors that send vital data directly by mobile phone to the emergency department of the closest receiving hospital. The system has been created to test a new strategy, devised by University of California-San Francisco (UCSF) researchers, that aims to speed treatment for heart attack patients.

In mountainous Santa Cruz County, ambulance runs are often lengthy. "Every minute that heart cells are deprived of blood flow, they are dying," says Barbara Drew, the study's principal investigator and a professor of physiological nursing in UCSF's School of Nursing. "Once heart cells are dead, they don't regenerate. So the initial treatment goal is to get the blockage in the obstructed artery open as quickly as possible before any more heart cells die."

The new "tele-electrocardiography" system consists of a cardiac monitor that takes readings every 30 seconds and can detect ischemia, the diminished flow of blood through an artery that signals heart damage. The unit is hooked to a mobile phone that transmits vital information directly to the emergency department of the receiving hospital. Drew based the device on her years of experience in cardiac intensive care unit. "Usually when patients arrive at a hospital, they are evaluated by a triage nurse," says Drew. "If their condition warrants it, they are attached to a cardiac monitor for further evaluation. But all that takes time. What we wanted to do was to move the clinical decision-making to a point before the patient even gets to the hospital."

The standard heart-monitoring procedure used by medics who respond to calls from people experiencing heart attack symptoms involves attaching a cardiac monitor with a single recording lead to the patient's chest. The monitor provides only a basic electrocardiogram (ECG) that measures the patient's heart rate and rhythm. The unit can't detect ischemia. Some ambulances in the United States are equipped with 12-lead cardiac monitors. Although these can detect ischemia, they require the attachment of 10 separate electrodes to the patient's chest and they make only a single 10-second recording, which may miss rapidly changing abnormalities common in heart attacks.

The "tele-electrocardiography" system Drew is using in her study consists of several components, including a 12-lead cardiac monitor that requires only five electrode attachments to the patient's chest. Another key component is software that analyzes the ECG every 30 seconds for signs of ischemia and heart damage. Once patients are attached to the monitor, ambulance medics push a button to send the first reading to an emergency department computer. An audible alert accompanies the transmission. If the software detects changes in subsequent ECGs, it automatically transmits them, as well.

Drew chose Santa Cruz County as a testing ground for the system because the county is large and mountainous and has only two hospitals, both located near the coast. For many residents, hospital transit times are long. "If hospital teams had advance notice of the patient's condition, it would give them time to get ready for immediate treatment," says Drew.

Drew and her researchers are now studying the impact that the mobile phone-delivered heart data is having on patients. Drew says it will take five years and the enrollment of hundreds of patients to determine whether the system is worth the cost. "It would be irresponsible to spend the money necessary to equip thousands of ambulances with the system unless it can be shown that, overall, outcomes such as better long-term health and survival are improved," says Drew.

Although telemedicine has enabled greater access to health services, the potential communication problems it brings could interfere with the technology's potential to improve the diagnosing and treating of illness.

Telemedicine is certainly valuable for delivering health- related services to remote areas, but the dynamics of the interactions associated with it can increase the likelihood of uncertainty, frustration, and unmet expectations for all involved, says Richard L. Street, a health communication authority at Texas A&M University.

Street, working with the Texas Tech University telemedicine program, has analyzed teleconsultations involving videoconferencing between a patient and primary care giver at one location and a specialist at another to identify patterns of talk that could affect quality of care. Although the teleconsultation may allow the specialist and primary care provider to exchange information and ideas, such a teleconsultation may restrict patient involvement in the encounter, he says. "While patients usually account for about 40 percent of the talk occurring in traditional consultations, they account for only 23 percent in teleconsultations," he notes.

Patients rarely asked questions or asserted a perspective or an opinion—something less than ideal considering how much a patient interacts with his or her doctor can have profound effects on diagnosis, treatment, and even health improvement, Street says. Patients who actively participate in consultations with their physicians, he explains, receive a greater amount of information, understand the issues better and are more satisfied with their care—all of which make for an overall improved quality of care.

Not only can patient participation affect the quality of a visit to the doctor, but a growing body of research indicates that it can contribute to improved health and healthier behavior, Street adds. Limited patient participation may be due to several factors, he notes.

The presence of an additional medical expert may unintentionally limit patient involvement as the two physicians converse with one another about the case, he says. Street also points to cultural and demographic variables, noting that patients who are more involved in their consultations tend to have more formal education and be in the middle to upper income bracket. However, people in remote, rural areas who are likely to have telemedicine encounters tend to be poorer and have less formal education compared with their urban and suburban counterparts.

Street also notes that very little group discussion takes place in these encounters, possibly due to the actual construct of the encounter, which places the patient and primary care giver side by side, facing a monitor with the specialist as their visual focal point. Additionally, linguistic differences play a factor. Street says clinicians often share a specialized linguistic code that allows them to better communicate but comes across as difficult to understand medical jargon to the patient.

Street recommends that the benefits of doctors being able to talk with each other can be further enhanced if they give patients more opportunities to speak by using partnership-building methods like asking for the patient's opinion and other patient-centered responses, such as offering encouragement and showing concern and interest in the patient. "These communication strategies would both legitimize and effectively increase patient participation in these encounters," he says.

Chapter 2

Nuts and Bits— Telecom Hardware, Software, and More

Advanced telecom services aren't of much use without sophisticated telecom hardware and software to make everything run. In the Telecosmos, phones are computers and computers are phones. Many communication devices sit on desks and tables; some are carried, and others are worn. Is that a lamp or a phone? Can this room's walls hear me? Why did my refrigerator just talk to me? In the years ahead, these won't necessarily be silly questions.

2.1 PERSONAL COMPUTERS

A PC doesn't look much like a telephone; however, when equipped with a microphone and speakers (or headphones), it sure acts like one. Plus, when it comes to sending and receiving e-mail and instant messages, it's hard to beat a PC. New chip and software developments will make PCs an even more useful communication tools.

To renew the venerable PC, vendors are starting at the heart of the matter— the processor. Back in 1981, the original IBM PC featured an amazingly modest—at least from today's perspective—4.77 MHz CPU. Twenty-three years later, the two leading PC processor makers—Intel and Advanced Micro Devices (AMD)—are relentlessly pushing processor speeds toward $4\,GHz$ on both desktop and laptop models.

Telecosmos: The Next Great Telecom Revolution, edited by John Edwards
ISBN 0-471-65533-3 Copyright © 2005 by John Wiley & Sons, Inc.

Yet raw speed isn't the only processor attribute that separates the latest PCs from their underpowered predecessors. New chip-oriented infrastructures, such as Intel's Hyper-Threading and AMD's Hyper-Transport, promise to give PC users added power and convenience beyond a processor's basic clock speed.

Hyper-Threading brings virtual parallel processing to a single CPU, allowing PCs to handle multiple tasks faster and without interruption. "You have one logical processor servicing whatever you're doing and one in background taking care of the maintenance tasks, such as virus scanning," says William Siu, general manager of Intel's desktop platforms group. Hyper-Threading debuted on the Pentium 4, and Windows XP and Linux both support Hyper-Threading. Although applications that take direct advantage of Hyper-Threading remain scarce, Intel claims that users running two standard CPU-intensive applications simultaneously can expect up to 25 percent faster execution.

AMD's HyperTransport, on the other hand, is a high-performance bus that allows a PC's key system components to communicate with each other at speeds up to 50 times faster than the PCI bus currently used in most PCs. "It's designed to increase the speed of communication between the integrated circuits in computers, telecom equipment, networking systems and so on," says Deepa Doraiswamy, a semiconductor industry analyst with Frost & Sullivan, a technology research company located in San Antonio, Texas. According to the HyperTransport Consortium, over 45 HyperTransport products are already available, including CPUs, security processors, core logic and bridge devices, IP cores, and test equipment.

Other significant PC architecture improvements include PCI Express (a faster and simpler serial-oriented version of the PCI bus that promises to reduce the size and cost of both plug-in cards and motherboards), Serial ATA (a high-speed storage interface that cuts down on the cabling within PCs), Serial-Attached SCSI (a speed-scalable and less power-hungry version of the familiar SCSI storage device interface that also allows the development smaller form factor drives), and ExpressCard (a new PC expansion card standard, based on PCI Express, that aims to replace PCMCIA cards with smaller, faster and cheaper plug-in modules).

Additionally, over the next couple of years, PC vendors will accelerate their transition from 32-bit to 64-bit technology, responding to enterprise customers who use powerful database and multimedia software and services. Already, 64-bit technology is appearing on high-end desktops from Dell Computer, Hewlett-Packard, and other vendors. "We're actually seeing 64-bit making pretty good inroads into the high-end content creation area, whether it be graphics or doing movies or videos," says Kevin Knox, director of worldwide business development for AMD.

2.1.1 Smaller and Smarter PCs

For years, analysts have predicted that sophisticated notebook PCs would eventually supplant desktop systems. Although that moment probably won't arrive for some time, the latest notebooks are certainly more powerful and easier to use than their predecessors. Vendors are working hard to make their notebook systems more like desktop PCs while preserving portability. IBM, for example, has developed a notebook that's influenced by origami, the Japanese art of paper folding. When the system—based on a standard ThinkPad T40 notebook—opens, the display automatically moves upward several inches for better viewing. As the display rises, the keyboard reflexively slides toward the user and rests at a desktop keyboard-like typing angle. "You can unfold the system if you're at a bigger space to get the benefits of a desktop PC and then refold it back up into the clamshell when you don't have the space," says Howard Locker, chief architect of IBM's personal computing division. The company hasn't yet set a release date for the notebook. "We're testing this [system] right now to see if people are willing to pay the extra cost," says Locker.

Many people are also looking at portable systems other than notebooks. Tablet PCs may be an alternative for many users, particularly those who need to work with large amounts of text or numeric data in mobile environments. The systems, which are designed to mimic the dimensions of a large paper notebook, include an operating system that lets users jot information with a pen-like stylus. Mike Stinson, vice president of mobile products for Gateway Computer, predicts that pen-based input will be "a requirement" for most portable system users by 2005. "It's just an easier way to take notes," he observes. Tablet PC shipments in the United States are set to climb from 260,000 units in 2003 to 2.25 million in 2005, according to statistics compiled by IDC, a technology research firm (and *CIO Magazine* sister company). State-of-the-art PDAs are also gaining traction in many enterprises, thanks to their low cost, small size, wireless connectivity, and miserly power consumption. Busy managers appreciate the devices' ability to handle a variety of simple tasks, ranging from e-mail to text entry.

Whether people rush toward notebooks and other mobile devices, a shift toward smaller PC forms seems inevitable. A growing number of enterprises are looking closely at thin-client devices, says Martin Reynolds, a fellow at research firm Gartner, located in Stamford, Connecticut. Thin clients, which link to a central server and have no internal disk storage, hold the promise of lower cost, better management, and enhanced security. "The market is more ready for them now," says Reynolds. Although the thin-client model has existed for many years, Reynolds believes that in a chaotic world of viruses, hackers, and seemingly endless "critical updates," it makes more sense than ever to manage systems at the server level rather than on individual desktops.

The PC world is rapidly marching from desktop- to device-based computing—an environment in which "technology is embedded in just about everything," says Chris Shipley, executive producer of The DEMO Conferences, a Menlo Park, California-based organization that showcases budding technologies. Shipley notes that networks will soon be "smart enough to know who you are and what sort of device you're connecting from—then they'll just scale the information appropriately for the device you're using."

2.2   HOME AUTOMATION

Integral to the Telecosmos is home automation and connectivity, which will allow people to gain greater control over their daily lives. Home automation and connectivity, however, do not just involve letting in a delivery person while away or turning on lights when on vacation. They are about managing the home ecosystem. With networking technologies and standards connecting the various areas of the home, the opportunity to extend these initiatives—and make the home "smarter"—is beginning to become a reality.

"What is happening to the home now happened to the enterprise in the 1990s," explains Erik Michielsen, a senior analyst at ABI, a technology research firm located in Oyster Bay, New York. "Supply chain management, enterprise resource planning, and customer relationship management were large initiatives that cut costs, enabled efficiencies, and drove revenue for service providers and cut customer costs. The home is not that different."

Within the home enterprise, residential wireless networking, high-speed Internet services, and smarter connected intelligent devices are reshaping the home by connecting security, entertainment, HVAC, lighting, and appliances in new ways. These trends are pushing traditional home automation markets closer to those traditionally occupied by the PC and consumer electronics industries. In short, the trends are creating new guidelines for the digital home and the applications available to the consumer. From 2003 to 2008, ABI estimates the home automation controls market will grow from $1.5 billion to over $3.8 billion, pushed in large part by mainstream consumer adoption.

As networking and connectivity continue to drive development of networked home automation and digital devices, more stakeholders have emerged. For years, the home automation and controls industry has been fragmented; however, a number of players in the market have changed product strategies and are now offering integrated home solutions or complementary networked solutions suites. The traditional automation companies continue to innovate, but many new entrants from established adjacent industries—including Samsung, Honeywell, Invensys, Motorola, Texas Instruments, Mitsubishi, and Philips—are looking to capitalize on new market opportunities in home automation.

Networking technologies such as Zigbee, Bluetooth, Wi-Fi, and Ultra Wide band are gaining momentum through consumer adoption, focused trials, or standards certification. Each will affect the home automation landscape by altering the way nodes connect and share information, thus altering consumer lifestyles and availability of choice. These technologies present numerous opportunities to network home control systems, such as lighting, security, HVAC, entertainment, and appliances.

On the entertainment front, a convergence of television, PC, and Internet technologies promises to provide even more television choices, if not superior programming. "Things like television and personal computers—I think that terminology will be as quaint as the 'Victrola' is today," says Jim Barry, a spokesperson for the Arlington, Virginia-based Consumer Electronics Association. Convergence means that it's likely that all or most audio and video programming will be sent through some version of the Internet by 2025. "It's very possible that you will get all your TV broadcasting over the Internet, but you'll still watch it on a big TV screen of some sort," says Ed Price, research director at the Georgia Institute of Technology's Interactive Media Technology Center.

Internet-based programming delivery, in addition to allowing viewers to call up shows on demand, will enable TV to be transformed into a two-way medium that lets users interact with programs. Home shopping networks, for example, will allow viewers to order products directly from the screen via touch-screen, remote control, or spoken commands. "Say you're watching *Friends*, and you want to buy that sweater Jennifer Aniston is wearing; you'll be able to hit a button and order it," says Barry.

Online shopping itself will be a much different and more buyer-friendly experience in 2025. Rather than staring at TV pitchpersons or catalogue-like Web pages, Virtual reality technology will allow viewers to examine products from various angles, even in 3-D. Haptic interfaces, which provide tactile feedback, will allow shoppers to touch and manipulate products as if they were actually examining the items inside a real-world store. Additionally, shoppers will not only be able to look at fashions but also see how they themselves would actually look wearing the items, thanks to stored personal 3-D images and dimension profiles.

2.3 WEARABLE COMPUTERS

It seems hard to believe that today's mobile phones or PDAs will get any smaller, given the fact that many of today's models are small enough to fit into a shirt pocket. But by 2025, phone circuitry may be even smaller and cheap enough to be integrated into clothing, perhaps inside a button or even woven into the fabric itself. "I can't wait for the day when I have a device that's small enough to fit into my pocket, which isn't true for most PDAs today," says Wim Sweldens, algorithms research department director at Bell Labs in Murray Hill, New Jersey.

In the Telecosmos, a convergence between mobile phones and PDAs is inevitable. By 2025, a converged phone-PDA would likely take the form of a wearable computer. Already widely used in industry to help repair technicians, inventory takers, and other workers who either can't or don't want to enter data into a handheld device, wearable computers will likely become a common fashion accessory by 2025. "We're going to allow technology more intimately into our lives . . . wearing them like a wedding ring, a watch, your glasses," says Astro Teller, CEO of BodyMedia, a Pittsburgh-based wearable computer manufacturer.

Figure 2-1 Lightweight, wireless computer.

Wearable computers have already revolutionized communications in many fields, including firefighting and emergency medical services, where information must flow fast in adverse work environments. George Elvin, a professor of architecture at the University of Illinois at Urbana-Champaign, thinks that lightweight, wireless computers may similarly transform the construction industry in the not-too-distant future (Fig. 2-1). "Building design and construction has been called the world's largest industry," says Elvin. "It is also one of the most inefficient."

Consensus estimates suggest that as much as 30 percent of building project costs are wasted through poor management of the design-construction process, says Elvin. "This waste represents more than $10 billion in the United States every year that could be directed toward improved design, better materials, and related improvements to our built environment." In an effort to improve building industry efficiency, Elvin recently led a study that examined the effects of using wireless-enabled portable computers to complete integrated design-construction projects. The study looked at systems that can be strapped to a toolbelt as well as pen-based electronic tablets.

Elvin says the study aimed to "measure the accuracy, timeliness, completeness, and efficiency of information exchange enabled by wearable computers." The study was based on interviews with architects and contractors, construction-site observations, and data from controlled experiments at the Illinois Building Research Council. In those experiments, three small structures were built using different communications devices: traditional paper documents, a pen-based tablet computer, and a wearable computer with flat-panel display.

"Results indicated that tablet and wearable computers may significantly reduce rework, while productivity decreased slightly when tablet and wearable computers were used," Elvin says. With paper documents, for example, 4.15 percent of total project time was spent redoing some aspect of the project, compared with 1.38 percent with the wearable computer. Elvin says communications that use paper were probably less efficient because the quality of paper documents faxed to job sites is often poor; electronic tablets or wearable computers, however, allow construction-team members to enlarge parts of documents to view greater detail.

Elvin says a dip of less than 8 percent in productivity indicated in the study "is typical of the initial decline in productivity observed when a new technology is introduced to a workforce in any field." Further study is needed to determine the long-term productivity impacts of tablet and wearable computers once the user had become proficient in their use."

2.4 SMART FABRICS

A mobile phone with lapels? An MP3 player with a zipper? In the world of "smart fabrics," clothing and electronics can be indistinguishable. Werner Weber, senior director of corporate research at Infineon Technologies, a Munich-based semiconductor design firm, believes that many of the gadgets people currently take for granted—including phones, home entertainment devices, health monitors, and security systems—will literally be woven into the fabrics they wear, walk over, and sit on. The Munich-based chipmaker recently developed a carpet that can detect the presence of people—guests or intruders—and then automatically activate a security system or light the way to an exit in the event of a fire or other emergency. The carpet is woven with conductive fibers with pressure, temperature, and vibration sensor chips, as well as LEDs, embedded into the fabric. "The goal is to present security services and guiding functions in buildings, such as hotels and airports," says Weber. "The first products will reach the market in two to three years," he predicts.

With smart fabrics, the material is the device, says Sundaresan Jayaraman, professor of polymer, textile, and fiber engineering at the Georgia Institute of Technology. "You'll never forget your mobile phone or PDA; it will be a part of the shirt you wear." Jayaraman is the inventor of a "Smart T-shirt" that uses optical and conductive fibers to detect bullet wounds and monitor the wearer's vital signs, such as heart rate and breathing. Jayaraman, who has been engaged in smart fabric research since 1996, observes that the technology has various applications, including for military personnel, law enforcement officers, astronauts, infants, and elderly people living alone. A commercial version of the shirt is available from New York-based Sensatex, which sells the garment to athletes and other people who want to monitor biometric data, such as heart rate, respiration rate, body temperature, and caloric burn. Information generated by the shirt is wirelessly transmitted to a personal computer and, ultimately, the Internet, where a coach, doctor, or other conditioning expert can examine the information. The shirt's wearers can access the data via a wristwatch, a PDA, or voice output.

2.5 EMBEDDED SYSTEMS

Within a decade or so, information access terminals will be everywhere, although they won't look like today's phones or computers. They may, for example, look like a Coke machine. Future soda dispensers will be linked—like almost everything else—to the Internet. Beverage prices may be raised or lowered in accordance with customer demand, sales promotions, or even the outside temperature. Likewise, home appliances, office equipment, automobiles, and perhaps even disposable items such as party hats and roller skates may feature on-board computing and telecommunications capabilities. The concept of inescapable computing is known as pervasive computing.

Embedded computers are already an integral part of modern life. They're increasingly becoming the brains behind the core mechanisms inside a variety of common products, including wireless devices, cars, automated elevators, climate control systems, traffic signals, and washing machines.

"Some experts estimate that each individual in a developed nation may unknowingly use more than 100 embedded computers daily," says Sandeep Shukla, an assistant professor of electrical and computer engineering at Virginia Tech. Shukla recently received a $400,000 grant from the National Science Foundation to help solve the problem of transitioning businesses and people from a world of desktop and handheld computers to embedded devices.

There are two performance factors critical to embedded computers: speed and quality of service. "If the power supplied by the battery is too low, the computer's performance is reduced," Shukla says. "The question is whether a compromise between performance and power is reasonable for a particular device or application."

Shukla wants to support the current and future uses of embedded computers by developing a power usage strategy that can guarantee maximum performance. This entails analyzing the complex probabilities of when computers will require power and how much power they will use. "It's similar to designing a network of traffic lights for a particular traffic pattern," he says. "The highway engineer has to study the probabilities of when and where traffic is the heaviest and then set up a network of lights that will allow a maximum flow of traffic."

One possible usage strategy would be to place a mobile phone into "sleep" mode during times when the probability of usage is low. The design would keep the system in a "ready" mode when incoming and outgoing calls are expected and fast action is required. Such a strategy would reduce power use and increase the life of the battery while optimizing the cell phone's performance.

Using a probability analysis modeling tool called PRISM, which he worked with at the University of Birmingham in England, Shukla plans to devise usage strategies for a network of wireless computers. By analyzing usage frequencies and probabilities of all the computers in a networked embedded system, Shukla hopes to create a strategy that will reduce power use while increasing performance. "Eventually, companies will use probability design in developing embedded computers for everything from small wireless devices to large-scale computer networks," says Shukla.

Shulka also plans to develop graduate and undergraduate courses in embedded computer systems and to support the work of student assistants in a new research laboratory he has founded.

2.6 PROJECT OXYGEN

Imagine a world where computers are everywhere and finding someone on a network will be as easy as typing or saying, "Get me Jane Doe at XYZ Corp. in Topeka, Kansas." That's the goal of Project Oxygen, an ambitious venture launched by the Massachusetts Institute of Technology that aims to make computing and electronic communication as pervasive and free as the air.

First proposed in 1999, Project Oxygen was the brainchild of Michael Dertouzos, the late director of MIT's renowned Laboratory for Computer Science. Dertouzos had a vision for replacing the PC with a ubiquitous—often invisible—computing and communications infrastructure. Today, Project Oxygen consists of 30 MIT faculty members who work with two MIT departments, the federal government, and several major technology companies in an effort to make information and communications access as easy to use and omnipresent as a light switch.

2.6.1 The Vision

Today's computing and communications systems are high-tech bullies. Rather than fitting into users' lifestyles, they force people to adapt themselves to the technology. Project Oxygen is designed to turn the status quo on its head by making information and communications access a natural part of everyday life.

The envisioned Project Oxygen system consists of a global web of personal handheld devices; stationary devices in offices, homes, and vehicles; and a dynamically configurable network. A related project seeks to design a new type of microchip that can be automatically reprogrammed for different tasks: this chip would power Oxygen devices.

MIT's role in Project Oxygen is to unite engineers, software developers, and other global computer experts to create a pervasive computer environment. "We find ourselves in the junction of two interrelated challenges: Going after the best, most exciting forefront technology; and ensuring that it truly serves human needs," wrote Dertouzos in a mission paper, shortly before his death.

Project Oxygen officially got underway in June 2000, when MIT formed a five-year, $50 million Project Oxygen Alliance with the Defense Department's Defense Advanced Research Projects Agency (DARPA) and several leading technology companies. Hewlett-Packard, Japan's Nippon Telegraph and Telephone, Finland's Nokia, the Netherland's Philips, and Taiwan's Acer and Delta Electronics are all working on key parts of the project Oxygen infrastructure. The companies will contribute $30 million by 2005, with the rest of the budget coming from DARPA. Two MIT labs are sharing responsibility for Project Oxygen: the Lab for Computer Science and the Artificial Intelligence Lab.

2.6.2 Goals

To succeed, Project Oxygen must meet four distinct goals, each a critical piece in the venture's overall structure. Once achieved, the goals will fulfill Project Oxygen's mission of bringing abundant and intuitive computation and communication tools to users.

The first goal, and the one perhaps most important, is pervasive computing and communication. The Oxygen system must be everywhere, with every portal reaching into the same information base. No longer will technologies, service providers, or geopolitical borders segregate users.

The project's next goal is to develop hardware and software tools that are embedded into users' daily lives. The researchers maintain that Oxygen's technology must live in the real world, sensing it and affecting it. Users shouldn't have to learn how to use the system; the system should be able to automatically adapt itself to its users' needs. Natural, perceptive interfaces, including voice and facial recognition and realistic graphics, will make it easy for people to perform tasks. Just as people don't have to read a thick instruction manual in order to turn on a new desk lamp, Oxygen users won't have to plow through pages of detailed and cryptic information whenever they acquire a new piece of technology.

"Nomadic computing"—the ability to access information and people anytime, any place—is another key Project Oxygen goal. Easy roaming will allow people to move around according to their needs instead of placing themselves at specific locations in order to handle information-related tasks.

Finally, the Oxygen environment must be eternal. Like power or phone services, the system should never shut down. While individual devices and software components may come and go in response to glitches and upgrades, the Oxygen system as a whole must operate nonstop and forever.

Project Oxygen relies on an infrastructure of mobile and stationary devices that are linked together by an intelligent, self-configuring network. The network will sit above the Internet and automatically adapt itself, depending on what device a user needs and the individual's location anywhere in the world.

Unlike conventional computers and mobile devices, which rely on keyboard, mouse, and touch input, Project Oxygen will use highly accurate speech recognition technology for more natural system interaction. Down the road, Oxygen researchers are planning to add vision-augmented speech recognition that will allow Oxygen devices to understand a user's intentions by recognizing facial expressions, lip movements, and even a user's gaze.

Project Oxygen's software environment will be designed to accommodate rapid changes, in both technology and user needs. The system will be able to absorb new features and specifications without affecting people using previous-generation devices. Customized software will play a key role in Oxygen's day-to-day use. An accountant, for example, would use one type of software, whereas a doctor would use another kind of programming package. A user could find himself or herself using several different types of Oxygen software sets: at work, at home, and at play.

An intelligent network, dubbed Network21 (N21), lies at the heart of Project Oxygen's communications infrastructure. The network will link an array of stationary and mobile devices. Besides providing communications links across cities, nations, and continents, N21 will support multiple communication protocols to provide low-power point-to-point, building-wide, and campus-wide communication.

2.6.3 User Technologies

Project Oxygen's user technologies will mark a radical departure from today's world of desktop PCs, laptop computers, and PDAs. By allowing users to seamlessly transition between stationary and mobile devices—without the need for time-consuming data syncing—Oxygen aims to make computing and communications almost effortless.

Because mobility is key to the Project Oxygen philosophy, handheld devices will be one of the system's most important—and interesting—technologies. The basic Oxygen mobile device is the Handy 21 (H21). This small, handheld unit will combine the features of mobile phones, pagers, portable computers, radios, TVs, and remote controls. Oxygen's developers envision a pocket-sized device that will incorporate a microphone, speaker, video screen, and camera. A global positioning system (GPS) module, which would allow the Oxygen system to pinpoint a user's exact location, will also be included.

Although H21s will serve as all-purpose, go-anywhere personal computing /communication devices, Project Oxygen researchers also want to bring homes and workplaces into the pervasive computing loop. Enviro21 (E21) devices—stationary units that feature an array of sensors, cameras, and micro-phones—will gather and transmit audio, video, and data information to users anywhere in the world. E21s will also allow users to access various types of information and communications resources and to control the local environ-ment. Users will be able to communicate naturally in the spaces created by E21s, via speech and vision interfaces, without relying on any particular point of interaction (such as a PC or telephone).

Unlimited by size, weight, power, or wireless connections, E21s will provide far more computational power than H21s. Oxygen's researchers believe the extra power will pay big dividends in terms of speech and facial recognition and other types of natural user interactions. Additionally, H21 users will be able to connect to a nearby E21 to access the device's abundant computational power.

2.6.4 Applications

Potentially equal to Project Oxygen's information and communications access capabilities are the ways this system will allow people to use information. Oxygen's applications promise to create a world where information flows as freely as water.

An obvious use of Oxygen's natural interface, communication, and control capabilities will be home and workplace automation. Users will be able to ver-bally create command sequences for controlling devices such as lights, doors, and heating and cooling systems. Want to raise the volume on your TV? It could be as simple as shouting, "Louder, please." Want to turn on your office coffee maker while you're driving into work? Simply bark the command into your H21.

Given Oxygen's anytime, anyplace audio/video delivery capabilities, home and workplace monitoring should be a snap. Anxious parents will be able to surreptitiously monitor a baby-sitter via their H21 while sitting in a movie theater or riding in a car. A boss could snoop on workers while sitting in his or her office or while attending a meeting in another country. Factory workers could monitor critical meters and gauges without tying themselves to a central control panel. Home patient monitoring is another potential Oxygen application.

Project Oxygen will also allow people to collaborate with each other in new and innovative ways. The H21, for example, will enable users to record and save highlights from meetings and speeches for future access. Videoconferencing could become commonplace as E21 systems are installed in a growing number of homes and workplaces. H21s would allow users to join a videoconference from almost anywhere, such as an airport departure lounge or from the backseat of a car. Oxygen's built-in speech and facial recognition technology will automatically identify conference participants and track each member's contributions to the proceedings.

Finally, Oxygen's impressive access capabilities will allow users to create their own custom knowledge bases. Like today's Web portals, only much more comprehensive and easier to use, Oxygen-powered knowledge bases will provide in-depth information on a particular topic or series of topics. Accessible by voice, and including multimedia content, a knowledge base will be able to collect material automatically, directed with basic commands from its operator. People will also be able to access knowledge bases operated by friends, business associates, and organizations worldwide. MIT researchers are also developing an advanced software technology that will organize information not only by structure, but by meaning.

2.6.5 Hurdles

Although few can argue with Project Oxygen's ultimate objective of creating a pervasive, natural computing, and communications environment, developing the underlying technology will be a remarkable achievement requiring plenty of hard work and numerous technological breakthroughs. The process will also require corporate hardware and software developers to closely cooperate on designs and standards, a process that doesn't come naturally to die-hard competitors.

The first hurdle in bringing Oxygen to fruition lies in creating hardware that's adaptable, scalable, and stream efficient. Researchers will also have to create software and protocols that are adaptable, flexible and intercompatible. Next in line will be the development of services and software objects that have names, not numbers, which will make the Oxygen environment easy for people to use. Also on the menu is software that is continuously operating yet replaceable on the fly, freeing software from hardware restraints.

None of these developments will come easy. Voice-recognition technology, for example, has followed a long and tortuous development road over the past several decades. Similarly, the amount of battery power required by the H21 doesn't yet exist. Battery technology, unfortunately, has advanced only incrementally over the past several years and no major breakthroughs are on the horizon.

Worse yet, even if existing technological barriers can be overcome, cost may prove to be Oxygen's ultimate undoing. For the system to become truly pervasive, it must be affordable to people in all segments of society. Right now, many of Oxygen's leading-edge technologies are priced far beyond the reach of average consumers.

MIT's researchers, however, remain undaunted. They are continuing to work on an array of Oxygen-related technologies and are hoping to drive costs down to realistic levels. The venture's corporate partners are also investigating key Oxygen hardware and software components as a part of their ongoing internal research and development efforts. The fruits of all this development work could begin showing up well before Project Oxygen's 2005 deadline, appearing on next-generation mobile phones and PDAs.

2.6.6 The Payoff

If everything goes according to plan, and Project Oxygen's various technological barriers are overcome, users can expect to see a vastly changed world. One of the venture's major benefits, as the technology makes it easier and cheaper for manufacturers to grind out vast quantities of identical products, will be the arrival of more efficient and less costly computing and communications technologies.

On the dark side, Oxygen is bound to raise privacy concerns. The Orwellian prospect of having microphones and cameras poking out of every corner will certainly discomfort more than a few people. Security could also become a major issue, with hackers potentially breaking into the system to spy on users and steal information.

Yet MIT maintains that Oxygen, over the long run, will be secure and will lead to more satisfied and productive computer users. If the school and its partners can bring the general public over to its side, Oxygen could turn out to be the great technology milestone of the 21st century.

2.7 THE OBJE SOFTWARE ARCHITECTURE

As the telecom world becomes increasingly complex and interconnected, imagine a platform that would allow people and businesses to access and deliver information and services from anywhere, on any device, in a completely hassle-free, ad hoc manner. Such a platform would dispose of the need to load device drivers and the need to worry about compatibility issues or complicated configurations. Xerox's Palo Alto Research Center (PARC) believes it has just such a technology with its Obje software, which uses mobile code (such as Java) to enable devices to "teach" each other how to interoperate in a user-friendly way.

The Obje software architecture is an interconnection technology that aims to allow digital devices and services to easily interoperate over both wired and wireless networks. At the architecture's heart is a simple "meta standard" for interoperation that allows users to access information and services from anywhere, on an ad hoc basis.

By providing a uniform solution to interoperation, the Obje platform is designed to make it easier for telecom vendors to build devices and services that work together. Putting assembly control into the hands of end users also reduces the burden of developing applications because particular customization can be performed in context.

Obje supports all standards, even those that have not yet been defined. The platform requires no central coordination, preconfiguring, or special setup and can be used by people with no technical expertise. It enables users to combine devices and build simple solutions, easily assembling applications from available devices and services. The platform offers device manufacturers a simple and fast solution to the growing need to connect products. Obje works with devices of all kinds, including mobile phones, computers, PDAs, printers, set top boxes, bar-code scanners, and video displays, and from any manufacturer.

Obje is designed to cut through complex protocols. Typically, communication among devices or services is structured into many protocol layers. Agreement on all layers is required before the devices and services are built. Developing and gaining acceptance of these agreements is a long, costly process that depends on broad industry consensus. Instead of working out all agreements in advance, Obje specifies a few very general agreements in the form of domain-independent programmatic meta-interfaces. The meta-interfaces use mobile code to allow new agreements to be put in place at run-time, enabling devices and services to dynamically extend the capabilities of their clients. The Obje meta-interfaces reduce the number of agreements that must be made between communicating entities. All Obje devices or services, called "components," implement and make use of one or more meta-interfaces.

PARC researchers have developed a variety of components and applications that use the architecture to cope with diverse performance, security, and usability requirements, as well as a variety of data types. Applications include a multimedia set top box, a public display system, and a system called "Casca," which allows team members to share documents and device resources such as cameras, printers, and speakers. Although Casca was designed to be a collaborative tool, no component functionality was hardwired into it. For example, Casca was not specifically written to support video conferencing, but it could acquire that functionality as soon as members of the group shared cameras, speakers, and microphones.

Obje is a key element of PARC's vision of ubiquitous computing, in which people are able to connect with the computers and telecom services that surround them, no matter where they are or what type of device they are using. It overcomes the problem of multiple, incompatible standards that prevents ubiquitous computing from becoming a reality. PARC is currently seeking corporate partners interested in using Obje inside their own products and applications.

2.8 BARN OPENS THE DOOR

New technology created by Carnegie Mellon University researchers will allow enterprises to create "smart rooms" that allow employees to participate in interactive electronic meetings, store important computer files, and secure sensitive research data.

The school's new BARN project provides the digital equivalent of dedicated meeting rooms. The technology is designed to give everyone in an organization, from entry-level clerks to upper management, the ability to instantaneously access a wide array of information from almost anywhere. Instead of seeping out over months and years, ideas can be zapped from an interactive project room to counterparts around the globe in a blink of an eye.

"The use of BARN technology will help companies and organizations become more fluid and molecular," says Asim Smailagic, a principal research scientist with Carnegie Mellon's Institute for Complex Engineered Systems (ICES). "Our powerful BARN tools permit users to organize, retrieve, store, and share information from multiple modes of collaboration," says Smailagic.

Companies using BARN will be able to perform as autonomous business units that are connected across geographies via a network. Every room using the technology will be seamlessly connected, allowing employees to work together in real time, knowing that confidential data can be secured at a moment's notice. Specialized interactive devices will direct all work done through BARN installations. Smailagic predicts that the speed of actions, information, and deliberations will increase as more companies adopt the technology.

The new technology includes software that allows users to access their digital files at once with a single interactive device. BARN also uses specially designed computer boards, remote interactive devices, and sophisticated 3-D audio systems for improved presentation of meeting information and knowledge transfer. "This project supports the nomadic character of today's business environment where mobile extension supports remote collaboration," says Smailagic. "Through the use of increased digitization, BARN allows companies and organizations, when a project demands, to replace minds and hands with computer networks to complete a task."

Group communication technologies are gaining greater visibility, particularly in places like Hong Kong and Singapore, where business trade had been off substantially because of the deadly severe acute respiratory syndrome (SARS) virus. Public health officials believe that SARS is spread by close contact between people; as a result, executives are now seeking new ways to conduct business, including the increased use of teleconferencing. Technologies like BARN may play an important role in a business world where threats posed by disease, war, and terrorists, could limit worker mobility.

2.9 PHONE AWARENESS

Future cellular telephones and other wireless communication devices are expected to be much more versatile as consumers gain the ability to program them in a variety of ways. Scientists and engineers at the National Institute of Standards and Technology (NIST) have teamed up with a variety of computing and telecommunications companies to develop both the test methods and the standard protocols needed to make this possible.

Programmable networks will include location-aware services that will allow users to choose a variety of "context aware" call-processing options depending on where they are and who they are with. For example, a cell phone that "knows" your location could be programmed to invoke an answering message service automatically whenever you are in a conference room or in your supervisor's presence. Context aware, programmable cell phone or PDA networks also may help users with functional tasks like finding the nearest bank or restaurant. Within organizations, these capabilities might be used to contact people by their role and location (e.g., call the cardiologist nearest to the emergency room).

Before such capabilities can be realized on common commercial systems, groundwork must be completed to design and test open specifications of features and rules and procedures for programmable call control systems and to develop protocols that will enable these systems to utilize context information. NIST, working with Sun Microsystems, has designed and developed new Java specifications (JAIN SIP) that provide a common platform for programmable communication devices. The NIST work is based on the Session Initiation Protocol, a specification for call control on the Internet. NIST's open source implementation (NIST SIP) is a prototype that serves as a development guide and facilitates interoperability testing by early industry adopters of this technology.

2.10 COGNITIVE SOFTWARE: ANTICIPATING USER INTENT

New "smart" software promises to fundamentally change the way people interact with computers, making it easier for users to express their intentions to an array of digital devices, including personal assistants and smart phones. The result could be machines that think and behave more like people.

Over the past five years, a research team at the U.S. Department of Energy's Sandia National Laboratories has been developing cognitive machines that accurately infer user intent, remember past experiences, and allow users to call on simulated experts to help them analyze situations and make decisions. "In the long term, the benefits from this effort are expected to include augmenting human effectiveness and embedding these cognitive models into systems . . . for better human-hardware interactions," says John Wagner, manager of Sandia's computational initiatives department.

The work's initial goal was to create a "synthetic human"—a program/computer—that could think like a person. "We had the massive computers that could compute the large amounts of data, but software that could realistically model how people think and make decisions was missing," says Chris Forsythe, a Sandia cognitive psychologist who is leading the research.

There were two significant problems with creating the original software. First, the software did not relate to how people actually make decisions. It followed logical processes, something people don't necessarily do. That's because humans often make decisions based on experiences and associative knowledge. Additionally, the model didn't take into account organic factors, such as emotions, stress, and fatigue, that are vital to realistically simulating the human thought processes.

Later, Forsythe developed the framework for a program that incorporated both cognition and organic factors. In follow-up projects, methodologies were developed that allowed the knowledge of a specific expert to be captured in the computer models as well as providing synthetic humans having memory of experiences. The approach allowed a computer to apply its knowledge of specific experiences to solving problems in a manner that closely paralleled what people do on a regular basis.

Forsythe says a strange twist occurred along the way. "I needed help with the software," he recalls. "I turned to some folks in robotics, bringing to their attention that we were developing computer models of human cognition." The robotics researchers immediately saw that the model could be used for intelligent machines, and the whole program emphasis changed. Suddenly the team was working on cognitive machines, not just synthetic humans.

The work on cognitive machines took off in 2002 with a contract from the Defense Advanced Research Projects Agency (DARPA) to develop a real-time machine that can infer an operator's cognitive processes. Early this year, work began on Sandia's Next Generation Intelligent Systems Grand Challenge project. "The goal of this Grand Challenge is to significantly improve the human capability to understand and solve national security problems, given the exponential growth of information and very complex environments," says Larry Ellis, the principal investigator. "We are integrating extraordinary perceptive techniques with cognitive systems to augment the capacity of analysts, engineers, war fighters, critical decision makers, scientists, and others in crucial jobs to detect and interpret meaningful patterns based on large volumes of data derived from diverse sources."

"Overall, these projects are developing technology to fundamentally change the nature of human-machine interactions," Forsythe says. "Our approach is to embed within the machine a highly realistic computer model of the cognitive processes that underlie human situation awareness and naturalistic decision making. Systems using this technology are tailored to a specific user, including the user's unique knowledge and understanding of the task."

The idea borrows from a very successful analogue. When people interact with one another, they modify what they say and don't say with regard to such things as what the person knows or doesn't know, shared experiences, and known sensitivities. The goal is to give machines highly realistic models of the same cognitive processes so that human-machine interactions have essential characteristics of human-human interactions.

"It's entirely possible that these cognitive machines could be incorporated into most computer systems produced within 10 years," Forsythe says.

2.11 DEVICES THAT UNDERSTAND YOU

A key step in the development of cognitive telecom products will be the addition of tiny sensors and transmitters to create a personal assistance link (PAL). With a PAL in place, a telecom device can become an anthroscope—an investigator of its user's vital signs. Such a system will monitor its user's perspiration and heartbeat, read facial expressions and head motions, and analyze voice tones. It will then correlate these factors to alert the user to a potential problem (such as talking too much during a phone call) and to anticipate questions instead of passively waiting for a request. The system will also transmit the information to other individuals within a virtual group so that everyone can work together more effectively.

"We're observing humans by using a lot of bandwidth across a broad spectrum of human activity," says Peter Merkle, a project manager at Sandia National Laboratories' Advanced Concepts Group. Merkle is using a Tom Clancy-based computer game, played jointly by four to six participants, to develop a baseline understanding of human response under stress (Fig. 2-2). "If someone's really excited during the game and that's correlated with poor performance, the machine might tell him to slow down via a pop-up message," says Merkle. "On the other hand, it might tell the team leader, 'Take Bill out of loop, we don't want him monitoring the space shuttle today. He's had too much coffee and too little sleep. Sally, though, is giving off the right signals to do a great job.'"

A recent study sponsored by Sandia indicates that personal sensor readings caused lower arousal states, improved teamwork, and better leadership in longer collaborations. The focus behind the effort, funded by Sandia's Laboratory-Directed Research and Development Program, is to map the characteristics that correlate to "personal-best" performances. "The question is, how do we correlate what we observe with optimum performance, so that we improve your ability and the ability of your team leader to make decisions? He can't tell, for example, that your pulse is racing. We're extending his ability," says Merkle.

Figure 2-2   Tom-Clancy based computer game.

People concerned about privacy—who may see this technology as an incursion similar to HAL's, the supercomputer that took over the spaceship in the movie *2001*—can always opt out, says Merkle, just like people choose not to respond to e-mails or decline to attend meetings. But in a sense the procedure is no different from that followed by people who have heart problems: they routinely wear a monitor home to keep informed of their vital signs. "In our game, what we learn from your vital signs can help you in the same way," he says. "It's almost absurd on its face to think you can't correlate physiological behavior with the day's competence."

No theory yet exists to explain why or how optimal group performances will be achieved through more extensive computer linkages. But Merkle doesn't think he needs one. "Some people think you have to start with a theory. Darwin didn't go with a theory. He went where his subjects were and started taking notes. Same here," he says. Further work is anticipated in joint projects between Sandia and the University of New Mexico and also with Caltech.

2.12   TURBOCHARGING DATA

A series of advances on the hardware component front promise to make telecom devices far more powerful than anything that's currently available. The technology that will drive future 4G networks is now being planned in research labs around the world. For example, a turbo decoder chip, developed by a Bell Labs research team in Sydney, Australia, offers the potential to speed data transmission on current 3 G and planned 4 G wireless networks. The chip handles data at rates of up to 24 Mbps—nearly 10 times faster than today's most advanced mobile networks.

The design opens up a new world of application possibilities for wireless handsets. According to Bell Labs, the device can support wireless TV-quality videoconferencing, multimedia e-mail attachments, and the quick downloading of high-quality music files. "This chip is a 'proof of concept'," says lead developer Mark Bickerstaff. "It demonstrates the possibility of receiving high-speed data on a handset."

The device was designed by the same team that recently announced the first chip incorporating Bell Labs Layered Space Time (BLAST) technology for mobile communications. The BLAST chip enables wireless devices to receive data at 19.2 Mbps in a 3 G mobile network. The team believes that the new chip's many commercial possibilities will make it an ultimately successful technology. "We believe the chip will inspire many enterprise applications because it will offer high-speed access to the office," Bickerstaff says.

The turbo decoder supports the evolving High Speed Downlink Packet Access (HSDPA) standard, an evolutionary enhancement to Universal Mobile Telecommunications System (UMTS) spread-spectrum technology, also known as wideband code division multiple access (W-CDMA). The chip is fast enough to support not only first-generation HSDPA systems, which will offer transmission speeds between 5 and 10 Mbps, but also future Multiple-Input/Multiple-Output (MIMO) HSDPA systems, which are expected to achieve peak data rates up to 20 Mbps.

The new chip marks an attempt by Lucent to raise carriers' interest in building faster wireless networks. Such networks would help sell cutting-edge wireless applications, such as video calls, which are projected to create $20 billion in total revenue by 2006, according to Gartner Dataquest, a Stamford, Connecticut-based technology research firm.

But don't expect to find turbocharged mobile phones on the market anytime soon. Precisely when the chip is actually used in mobile phones or other wireless devices depends on how fast wireless carriers roll out high-speed networks. In the United States and many other parts of the world, which could take years. Lucent predicts carriers will begin to use the new HSDPA standard by 2006, but that time frame could be sooner if the economy revives and customer demand drives the wireless companies to move faster. "The chip makes Lucent attractive to handset partners, and that broadens our image in the wireless industry," says Bickerstaff. "Now we're not only known for our wireless base stations, we're also becoming known for our chip designs for handsets as well."

The potential for further innovations looks very promising," says Chris Nicol, Bell Labs' wireless research technical manager. "We're creating the future, and we intend to stay out in front.

2.12.1 Faster Transistor

IBM researchers have developed a hybrid transistor that could mean speedier and far less power-hungry wireless devices. The transistor has the potential to improve wireless device performance by a factor of four or reduce power consumption by a factor of five.

Transistors are electrical switches that control the flow of current through chips. Boosting transistor efficiency means greater overall computer power, allowing mobile phones, PDAs, and other portable communications devices to handle more complex tasks, such as video streaming. Concurrently, lower power consumption leads to extended battery life for the same kinds of devices.

As the wireless industry expands, device manufacturers need better mixed-signal chips, combining digital and analog processing, to support both computing applications and high-frequency communications applications. The new chip design, developed by Ghavam Shahidi and colleagues at IBM's Thomas J. Watson Research Center in Yorktown Heights, New York, uses a novel type of wafer that's thin enough to maximize the performance of both the computing and communications components.

Complementary metal oxide semiconductor (CMOS) chips provide the current foundation for computing applications. Silicon germanium (SiGe) bipolar chips are commonly used to provide radio frequency communications and analog functions. To improve the reliability of wireless devices, chip manufacturers create SiGe BiCMOS chips that place computing and communications transistors onto one chip instead of using separate chips for computing and communications applications.

CMOS computing chips show higher performance when built on top of a thin silicon-on-insulator (SOI) wafer. However, traditional SiGe bipolar transistors cannot be built on a thin SOI wafer. Until IBM's breakthrough, no one had been able to find a technique to combine CMOS and SiGe bipolar onto one wafer that would maximize the performance of both. The IBM researchers are the first to build SiGe bipolar using a thin SOI wafer, thereby paving the way to build SiGe bipolar and CMOS on the same thin SOI wafer, maximizing the performance of both the computing and communications functions.

"As the wireless industry continues to grow, new devices will require greater functionalities, performance, and reliability from their components," says T. C. Chen, vice president of science and technology for IBM Research. "IBM continues to find new methods to improve chips to ensure that the industry can meet consumer demands. The new chip design could be implemented within five years, enabling applications such as video streaming on cell phones."

2.12.2 Cutting-Edge Manufacturing

Vendors would like to make mobile phones small enough to embed into everyday objects, such as jewelry, eyeglass frames, or even clothing. First, however, researchers need to find a way to pack more transistors onto chips.

Now, a new manufacturing technique, one that combines two chip manufacturing processes, promises to make tiny phones an everyday reality.

Lithography, the current technology chipmakers use to build the different layers that make up microelectronic devices, is cost prohibitive at a smaller scale. Another approach to building smaller microelectronic devices exists, yet has its own limitations. The alternative method relies on long chains of molecules, called block copolymers, which arrange themselves into patterns on a given surface. With self-assembling materials, creating exceptionally tiny circuits is inexpensive and routine. This method is inexpensive, but it is hampered by a high rate of defects and several other fabrication problems.

The new manufacturing technique combines lithography and self-assembly. By merging the two processes, researchers at the University of Wisconsin at Madison and the Paul Scherrer Institute in Switzerland developed a hybrid approach that maximizes the benefits and minimizes the limitations of each technique. "Our emphasis is on combining the approaches, using the desirable attributes of both, to get molecular-level control in the existing manufacturing processes," says Paul Nealey, a University of Wisconsin-Madison chemical engineer.

Specifically, the researchers used lithography to create patterns in the surface chemistry of a polymeric material. Then, they deposited a film of block copolymers on the surface, allowing the molecules to arrange themselves into the underlying pattern without imperfections "Tremendous promise exists for the development of hybrid technologies, such as this one in which self-assembling materials are integrated into existing manufacturing processes to deliver nanoscale control and meet exacting fabrication constraints," says Nealey.

About every 18 months, the number of transistors in computer chips doubles—the direct result of ever-shrinking sizes. By decreasing component size and, consequently, fitting more of them onto a single chip, computer speed and power improve. For this trend to continue during the next 20 years, science, technology, and the techniques used to produce microelectronics will need to operate on an even smaller scale. "Where the electronics industry is going is the manufacturing of devices that have ever-decreasing dimensions," says Nealey. He notes that the goal is to take the scale down from around 130 nanometers to under 50 nanometers (a nanometer equals one billionth of a meter).

The research was conducted at the Center for NanoTechnology at the University of Wisconsin at Madison's Synchrotron Radiation Center. It was funded in part by the National Science Foundation's Materials Research Science and Engineering Center and the Semiconductor Research Corp., a consortium that sponsors university research worldwide.

2.12.3 Wireless Chip

The silicon chip is destined to join the growing list of devices to go wireless, a development that could speed computers and lead to a new breed of useful products.

Figure 2-3 First wireless communication system built on a computer chip.

A team of researchers headed by a University of Florida electrical engineer has demonstrated the first wireless communication system built entirely on a computer chip (Fig. 2-3). Composed of a miniature radio transmitter and antenna, the tiny system broadcasts information across a fingernail-sized chip.

"Antennas are going to get installed onto chips one way or another—it's inevitable," says Kenneth O, a UF professor of electrical and computer engineering and the lead researcher. "We are really the first group that is making the technology happen." The major sponsor of O's five-year research project is Semiconductor Research Corp. (about $1 million).

As chips increase in size and complexity, transmitting information to all parts of the chip simultaneously through the many tiny wires embedded in the silicon platform becomes more difficult, O says. Chip-based wireless radios could bypass these wires, ensuring continued performance improvements in the larger chips. These tiny radios-on-a-chip could also make possible tiny, inexpensive microphones, motion detectors, and other devices, O says.

The fastest chips on the market—used in the Pentium 4 and other high-end processors—now operate at a speed of 2 GHz, meaning they perform 2 billion calculations per second, O says. Manufacturers are rapidly developing techniques to raise the speed, with chips that process information as fast as 20 GHz, or 20 billion calculations per second, already achieved on an experimental basis, he says. Many experts believe even 100-GHz chips are feasible.

The increase in speed will be accompanied by an increase in chip size, O says. Whereas today's average chip is about $1 cm^2$, or slightly under 0.5 inch, the faster chips anticipated in the next two decades are expected to be as large as 2 or 3 cm, or about 1.2 inches, on each side, he says.

The larger the chip, the harder it is to send information to all of its regions simultaneously because the distances between the millions of tiny circuits within the chip become more varied, O says. This can impact the chip's performance when the delay affects distribution of the so called "clock signal," a basic signal that synchronizes the many different information-processing tasks assigned to the chip. For optimum performance, this signal must reach all regions of the chip at essentially the same time. O and his colleagues have recently broadcast this clock signal from a tiny transmitter on one side of a chip a distance of 5.6 mm, or about a fifth of an inch, across the chip to a tiny receiver at the other end, avoiding all wires within the chip itself.

"Instead of running the signal through the wires, what we did was broadcast and receive the signal," O says. The demonstration proved it is possible to use a wireless system to broadcast on-chip signals.

The potential applications for chip-based radios go beyond maintaining the performance of larger chips, O says. In general, the availability of such chips could lead to a chip-to-chip wireless communication infrastructure, seamlessly and constantly connecting desktops, handheld computers, mobile phones, and other portable devices. The military has expressed interest in pairing wireless chips with tiny sensors such as microphones. The idea is to drop thousands or even hundreds of thousands of these devices in a region to eavesdrop over a wide area. The chips would form a listening network by themselves, and the military could monitor the system as needed, O says. On the civilian side, O says, scientists and engineers have theorized that the wireless chips could be paired with motion detectors and implanted in the walls of buildings. If a building collapsed due to an earthquake, for instance, the network of chips could broadcast information about movement to rescuers in search of victims.

2.12.4   Open Source Smart Phones

Symbian, Microsoft, and PalmSource dominate the smart phone operating system landscape. But, as in the PC field, open-source software could soon become a significant player in the budding smart phone operating system market.

The smart phone industry recently took a major step toward embracing open-source software when Motorola became the first major mobile phone maker to base a handset on Linux. The Motorola A760 is designed to combine the features of a mobile phone with the capabilities of a PDA, digital camera, and video/MP3 player. The product runs Java and related software from Sun Microsystems. The device also offers a variety of messaging functions, a speakerphone, Internet access, and Bluetooth wireless technology. On the hardware side, the A760 includes a color touch-screen and integrated camera.

Unlike proprietary smart phone operating systems, Linux is available to handset makers at reduced or even at no cost. But the software's price tag isn't a major consideration for Motorola. Instead, the Schaumburg, Illinois-based company is looking to tap into the rich supply of software available from open-source developers. "This handset is special because it features one of the most open and flexible software platforms that exists," says Rob Shaddock, a Motorola vice president and general manager of the GSM/TDMA product line in Motorola's personal communications unit. "By supporting the open-source Linux OS and Java technology, Motorola is creating the most open and flexible environment possible to help drive the development of compelling applications for rich, customized mobile experiences."

Smart phone industry observers and players are following Motorola's Linux handset with interest. But Neil Strother, an analyst with In-Stat/MDR, a technology research company based in Scottsdale, Arizona, doesn't believe that Linux-based smart phones are likely to snare a large market share, even five years down the road. "They'll probably grab 0 to 5 percent of the market," he says.

Still, given the likely prospect of massive smart phone sales—Strother feels that 49 percent of all mobile phones could be smart phones by 2008—even shipments in the low single digits could make Linux a factor to deal with. "Linux has a play because it's affordable, easy to use and deploy, and its features really do matter to the user," says Strother. Five years from now, the smart phone operating system field should be divided between Symbian at 53 percent, Microsoft at 27 percent, PalmSource at 10 percent, and Linux at about 4.2 percent, according to figures from IDC, a technology research firm located in Framingham, Massachusetts. Still, even if Linux sales don't soar, Motorola can take some comfort in the fact that it's also a member of the Symbian consortium.

Strother feels that the biggest immediate potential for Linux-based smart phones lies in Asia, particularly in China, where government authorities strongly favor nonproprietary software. "Motorola's idea is that, if the Chinese are going with open-source for servers and other corporate and government systems, then it makes sense for them to come up with handsets that are Linux-based."

2.12.5 Nanowiring

A new generation of cheaper, lighter, and more powerful mobile phones, PDAs, and other mobile devices could arrive within a decade, thanks to a nanotechnology breakthrough made by Harvard University researchers.

The researchers recently demonstrated that they can easily apply a film of tiny, high-performance silicon nanowires to glass and plastic. The development could lead to an array of futuristic products, including disposable computers and optical displays that can be worn in clothes or contact lenses.

Currently, amorphous silicon and polycrystalline silicon are considered to be the state of the art materials for making electronic components such as computer chips and LCDs; however, silicon nanowires are considered even better at carrying an electrical charge. Although a single nanowire is one thousand times smaller than the width of a human hair, it can carry information up to 100 times faster than similar components used in current consumer and business electronic products.

Scientists have already demonstrated that silicon nanowires have the ability to serve as components of highly efficient computer chips and can emit light for brilliant multicolor optical displays. But they have had difficulty until now in applying these nanowires to everyday consumer products. "As with conventional high-quality semi-conducting materials, the growth of high-quality nanowires requires relatively high temperature," explains says Charles M. Lieber, head of the research project and a Harvard chemistry professor. "This temperature requirement has—up until now—limited the quality of electronics on plastics, which melt at such growth temperatures."

By using a "bottom-up" approach, involving the assembly of preformed nanoscale building blocks into functional devices, the researchers have been able to apply a film of nanowires to glass or plastics long after growth and to do so at room temperature. Using a liquid solution of the silicon nanowires, the researchers demonstrated that they can deposit the silicon onto glass or plastic surfaces—similar to applying the ink of a laser printer to a piece of paper—to make functional nanowire devices. They also showed that nanowires applied to plastic can be bent or deformed into various shapes without degrading performance, a plus for making the electronics more durable.

According to Lieber, the first devices made with the new nanowire technology will probably improve on existing devices such as smart cards and LCD displays, which utilize conventional amorphous silicon and organic semiconductors that are comparatively slow and already approaching their technological limitations. Within the next decade, consumers could see more exotic applications of this nanotechnology, Lieber says. "One could imagine, for instance, contact lenses with displays and miniature computers on them, so that you can experience a virtual tour of a new city as you walk around wearing them on your eyes, or alternatively harness this power to create a vision system that enables someone who has impaired vision or is blind to 'see'."

The military should also find practical use for this technology, says Lieber. One problem soldiers encounter is the tremendous weight—up to 100 pounds—that they carry in personal equipment, including electronic devices. "The light weight and durability of our plastic nanowire electronics could allow for advanced displays on robust, shock-resistant plastic that can withstand significant punishment while minimizing the weight a soldier carries," says Lieber.

Many challenges still lie ahead in nanowire research, such as configuring the wires for optimal performance and applying the wires over more diverse surfaces and larger areas. Lieber recently helped start a company, NanoSys, that is now developing nanowire technology and other nanotechnology products.

2.13 MEMS

A key technology that will impact all sorts of telecom products in the years ahead are micro-electro-mechanical Systems (MEMS). MEMS is the integration of mechanical elements, sensors, actuators, and electronics on a common silicon chip through microfabrication technology. MEMS promises to revolutionize nearly every telecom product category by bringing together silicon-based microelectronics with micromachining technology, making possible the realization of complete systems-on-a-chip. MEMS is an enabling technology, allowing the development of smart products, augmenting the computational ability of microelectronics with the perception and control capabilities of microsensors and microactuators, and expanding the space of possible designs and applications.

2.13.1 Low-Loss, Wide-Bandwidth MEMS

Microelectronics researchers at the University of Illinois have developed a low-loss, wide-bandwidth microelectromechanical systems (MEMS) switch that can be integrated with existing technologies for high-speed electronics. The new low-voltage switch could be used in switching networks for phased-array radars, multibeam satellite communications systems, and wireless applications. "The switch has a tiny metal pad that can move up or down in less than 25 microseconds," says Milton Feng, a professor of electrical and computer engineering at the University of Illinois. "This simple configuration provides a very low insertion loss of less than 0.1 dB, and the metal-to-metal contact has the inherently wide-band response of a larger, more typical mechanical switch."

The switches are fabricated at the University of Illinois' Micro and Nano-technology Laboratory using standard MEMS processing techniques. To create the unique metal pull-down pad, Feng and graduate students David Becher, Richard Chan, and Shyh-Chiang Shen first deposit a thin layer of gold on a sacrificial layer of photosensitive material. Then they dissolve the sub-strate, pick up the pad, and place it in position on the switch. The metal pad—about 150 μm wide and 200 μm long—is supported at the four corners by serpentine cantilevers, which allow mechanical movement up and down.

"When in the 'up' position, the metal pad forms a bridge that spans a segment of the coplanar waveguide and allows the signal to pass through," Feng says. "But an applied voltage will pull the pad down into contact with the signal line, creating a short circuit that blocks the signal transmission." The gap between the metal pull-down pad and the bottom electrode is about 3 μm wide, which provides an isolation of greater than 22 dB for signal frequencies up to 40 GHz. Currently, an activation voltage of 15 volts is required to operate the switch.

One major problem Feng and his students had to overcome was stiction—a tendency for the metal pad to stick to a dielectric layer beneath the bottom electrode as a result of accumulated electrostatic charge. To prevent the charge from building up, the researchers added a tiny post that limits the downward motion of the pad. "This hard stop prevents the pad from moving past the bottom electrode and contacting the dielectric," Feng says.

In reliability tests, the switches have demonstrated lifetimes in excess of 780 million switching cycles. To further enhance the reliability, the researchers are attempting to lower the actuation voltage to less than 10 volts. "For any device to be used in a practical application it must be reliable," Feng says. "Our results show that good reliability is possible with low-voltage operation."

2.13.2 StressedMetal MEMS

The StressedMetal MEMS process is a proprietary technology for building three-dimensional MEMS devices. It is one of several MEMS technologies under development at Xerox's Palo Alto Research Center (PARC).

Traditionally, creating three-dimensional MEMS structures requires elabo-rate deposition and etching processes to build structures layer upon layer. StressedMetal MEMS offer a simpler approach. The process takes advantage of the stress that occurs in the thin-film deposition process. In thin-film depo-sition, extremely thin layers of metal film are deposited onto a substrate, such as glass or amorphous or silicon. PARC scientists have developed techniques to precisely control the stress within the layers of deposited metal. Litho-graphic techniques are used to etch patterns into the film layers. These patterns allow the metals to be released from the substrate. This technique can be used to create high volumes of self-assembling three-dimensional structures.

Devices fabricated with the StressedMetal process offer significant cost and performance advantages over those created with more traditional approaches. StressedMetal devices can be cost-effectively created using surface micro-machining techniques and standard batch-manufacturing processes. They can be integrated on silicon substrates and may include fully active circuitry. Or designers can use less expensive substrates, such as glass. Such substrates offer the added advantage of being much larger than standard silicon wafers.

To create StressedMetal MEMS structures, films are sputter deposited with an engineered built-in stress gradient. After lithographic etching is completed, the structures are released. The metal's inherent stress causes it to lift or curl into a designed radius of curvature, creating three-dimensional structures such as tiny coils, springs, or claws.

PARC's StressedMetal MEMS have been used in a number of prototype applications. All use simple and conventional fabrication processes, are able to be produced with high reliability, and offer superior performance to conventional counterparts. For example, PARC's StressedMetal on-chip, out-of-plane inductor enables a new generation of radio frequency integrated circuits (RFIC) and microwave components. PARC StressedMetal coils self-assemble into three-dimensional scaffolds that, when electroplated with copper, form highly conductive coil windings, suitable for future cell phones, TV tuners, wireless local networks, and other devices.

High-quality coils are the last remaining circuit components that cannot be integrated on-chip. As a result, most radio frequency circuits are built with modular assemblies containing many discreet components. PARC Stressed-Metal coils can be fabricated directly on silicon or gallium arsenide, thus enabling these modular assemblies to be replaced with a single integrated component. For example, voltage-controlled oscillators (VCO), used in cellular phone circuit boards, are modular circuits made up of many discreet components. These coils could be used to integrate these into a single chip that would be less than one-fifth the size of existing circuits and cost less to produce. Higher levels of integration are also possible.

StressedMetal coils can be manufactured on a range of substrates, including fully active circuit wafers. They offer performance up to twice that of comparably priced alternatives. The process for creating StressedMetal MEMS structures is similar to the deposition techniques used throughout the semiconductor industry to grow thin film layers—ranging from less than 1 μm to tens of micrometers—on substrates. The process uses a tool called a magnetron sputter system to bombard a surface with individual metal atoms. Beneath the bottom layer of atoms is a "sacrificial layer," which anchors the thin film to the substrate. Stripes or cantilevers are etched into the layers of metal atoms to define the shapes of the desired structures. Sections of the sacrificial layer are then dissolved with a chemical etchant, freeing the metal film from the substrate in the places where the sacrificial layer has been dissolved.

To create StressedMetal MEMS, scientists change the deposition parameters for each layer of atoms. They create two to five layers of stress-engineered metal by depositing the atoms with varying amounts of energy. They use a high level of energy for the bottom layer, causing the atoms to be packed very tightly. In successive layers, they use progressively less energy, until, at the top layer, the atoms land with very little force, analogous to that of gently falling snow. This technique harnesses both tensile and compressive stress of metal. Pairs of metal atoms act as springs, causing them to push apart (compressive stress) or pull closer (tensile stress) to maintain a consistent distance between them.

Tensile stress is caused when the loosely spaced atoms at the top pull more tightly together, as their electron clouds overlap and bond to one another. Compressive stress is caused when the tightly packed atoms on the bottom layer expand and push away from each other. When the metal is freed from the sacrificial layer, the compressive and tensile stresses in each layer bend the metal into the prescribed shapes.

2.13.3 The Nanoguitar

Several years ago, Cornell University researchers built the world's smallest guitar—about the size of a red blood cell —to demonstrate the possibility of manufacturing tiny mechanical devices using techniques originally designed for building microelectronic circuits. Now, by "playing" a new, streamlined nanoguitar (Fig. 2-4), Cornell physicists are demonstrating how such devices could substitute for electronic circuit components to make circuits smaller, cheaper, and more energy efficient. The guitar is so small that it falls into the nano-electro-mechanical system (NEMS) family. NEMS usually refers to devices about two orders of magnitude smaller than MEMS.

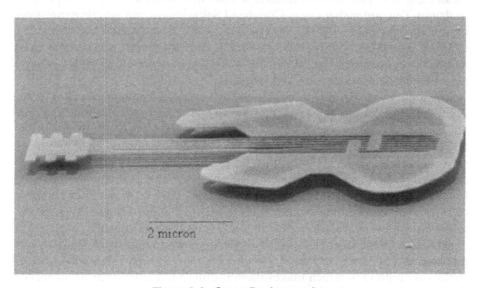

Figure 2-4 Streamlined nanoguitar.

Lidija Sekaric, now a researcher at IBM's Watson Research Center in Yorktown Heights, New York, worked with Cornell graduate student Keith Aubin and undergraduate researcher Jingqing Huang on the new nanoguitar, which is about five times larger than the original but still so small that its shape can only be seen in a microscope. Its strings are really silicon bars, 150 by 200 nm in cross-section and ranging from 6 to 12 µm in length (a micrometer is one-millionth of a meter; a nanometer is a billionth of a meter, the length of three silicon atoms in a row). The strings vibrate at frequencies 17 octaves higher than those of a real guitar, or about 130,000 times higher.

The researchers recently observed that light from a laser could cause properly designed small devices to oscillate, and this effect underlies the nanoguitar design. The nanoguitar is played by hitting the strings with a focused laser beam. When the strings vibrate, they create interference patterns in the light reflected back, which can be detected and electronically converted down to audible notes. The device can play only simple tones, although chords can be played by activating more than one string at a time. The pitches of the strings are determined by their length, not by their tension as in a normal guitar, but the group has "tuned" the resonances in similar devices by applying a direct current voltage.

"The generations of researchers to come can aim to play more complex pieces," says Sekaric. "This goal would indeed improve the science and technology of NEMS by aiming for integrated driving and detection schemes as well as a wide range of frequencies produced from a small set of vibrating elements."

Most of the devices the group studies don't resemble guitars, but the study of resonances often leads to musical analogies, and the natural designs of the small resonant systems often leads to shapes that look like harps, xylophones, or drums. The guitar shape was, Craighead Sekaric says, "an artistic expression by the engineering students." Sekaric notes that "a nanoguitar, as something close to almost everybody's understanding and experience, can also be used as a good educational tool about the field of nanotechnology, which indeed needs much public education and outreach."

The ability to make tiny things vibrate at very high frequencies offers many potential applications in electronics. From guitar strings on down, the frequency at which an object vibrates depends on its mass and dimensions. Nanoscale objects can be made to vibrate at radio frequencies (up to hundreds of megahertz) and so can substitute for other components in electronic circuits. Cell phones and other wireless devices, for example, usually use the oscillations of a quartz crystal to generate the carrier wave on which they transmit or to tune in an incoming signal. A tiny vibrating nanorod might do the same job in vastly less space, while drawing only milliwatts of power.

Research by the Cornell NEMS group has shown that these oscillations can be tuned to a very narrow range of frequencies—a property referred to in electronics as "high Q"—which makes them useful as filters to separate signals of different frequencies. They also may be used to detect vibrations to help locate objects or detect faint sounds that could predict the failure of machinery or structures.

As the nanoguitar shows, NEMS can be used to modulate light, meaning they might be used in fiber-optic communications systems. Such systems currently require a laser at each end for two-way communication. Instead, Craighead suggests that a powerful laser at one end could send a beam that would be modulated and reflected back by a far less expensive NEMS device. This could make it more economical to run fiber-optic connections to private homes or to desktop computers in an office.

2.14 STORAGE

As mobile devices become more capable, they'll need to store a growing amount of data. Getting tiny mobile units to store vast quantities of information isn't easy, however, given physical space restraints. But researchers are working hard to pack data into ever-smaller amounts of space.

2.14.1 Tiny Hard Drive

Toshiba has developed a 0.85-inch hard disk drive, the first hard drive to deliver multi-gigabyte data storage to a sub one-inch form factor. The device is suitable for use in a wide range of mobile devices, including palmtops, ultra-portable notebook PCs, handheld GPS units, and digital audio players and jukeboxes.

With the new drive, Toshiba has achieved a smaller, lighter, high-capacity storage medium in which low-power consumption is complemented by high performance. The drive will have an initial capacity of 2 to 4 GB and deliver enhanced data storage to smaller, lighter more efficient products. Toshiba expects the new drive to bring the functionality and versatility of hard disk drives to a wide range of devices, including mobile phones, digital camcorders, and external storage devices, as well as inspire other manufacturers to develop new applications. The device is scheduled to begin appearing in mobile devices during 2005.

Work on the drive has centered on Toshiba's Ome Operations-Digital Media Network, home to the company's main development site for digital and mobile products and the manufacturing site for the device. The drive under development is planned to have a capacity of 2 to 4 GB, but Toshiba anticipates achievement of even higher densities in the near future.

2.14.2 Optical Storage

A new optical storage medium, developed jointly by engineers at Princeton University and Hewlett-Packard, could profoundly affect the design and capabilities of future mobile devices, including mobile phones and PDAs.

The discovery of a previously unrecognized property of a commonly used conductive polymer plastic coating, combined with very thin-film, silicon-based electronics, is expected to lead to a memory device that's compact, inexpensive, and easy to produce. The breakthrough could result in a single-use memory card that permanently stores data and is faster and easier to use than a CD. The device could be very small because it would not involve moving parts such as the laser and motor drive required by CDs. "We are hybridizing," says Stephen Forrest, the Princeton electrical engineering professor who led the research group. "We are making a device that is organic—the plastic polymer—and inorganic—the thin-film silicon—at the same time."

The device would be like a CD in that writing data onto it makes permanent physical changes in the plastic and can be done only once. But it would also be like a conventional electronic memory chip because it would plug directly into an electronic circuit and would have no moving parts. "The device could probably be made cheaply enough that one-time use would be the best way to go," Forrest says.

Hewlett-Packard researcher Sven Möller made the basic discovery behind the device by experimenting with a polymer material called PEDOT, which is clear and conducts electricity. The material has been used for years as an anti-static coating on photographic film and more recently as an electrical contact on video displays that require light to pass through the circuitry. Möller found that PEDOT conducts electricity at low voltages but permanently loses its conductivity when exposed to higher voltages and currents, making it act like a fuse or circuit breaker.

This finding led the researchers to use PEDOT as a way of storing digital information. A PEDOT-based memory device would have a grid of circuits in which all the connections contain a PEDOT fuse. A high voltage could be applied to any of the contact points, blowing that particular fuse and leaving a mix of working and nonworking circuits. These open or closed connections would represent "zeros" and "ones" and would become permanently encoded in the device. A blown fuse would block current and be read as a "zero," whereas an unblown one would let current pass and serve as a "one."

The memory circuit grid could be made so small that, based on the test junctions the researchers made, 1 million bits of information could fit in a square millimeter of paper-thin material. If formed as a block, the device could store more than one gigabyte of information, or about 1,000 high-quality images, in one cubic centimeter, which is about the size of a fingertip. Developing the invention into a commercially viable product will require additional work on creating a large-scale manufacturing process and ensuring compatibility with existing electronic hardware, a process that might take as few as five years, Forrest says.

The technology offers numerous potential mobile device applications. Extensive and detailed street map databases, designed for use with GPS and other location-oriented services, could be easily inserted into even the smallest mobile devices and consume very little power. Other possible applications include easily accessible music and e-book libraries, shopping and attraction directories, and powerful software applications.

Funding for Forrest's research came in part from Hewlett-Packard as well as from the National Science Foundation. Princeton University has filed for a patent on the invention. Hewlett-Packard has an option to license rights to the technology.

2.14.3 Nanoring Memory

Recent nanotechnology research at Purdue University could pave the way toward faster computer memories and higher density magnetic data storage, all with an affordable price tag.

Just like the electronics industry, the data storage industry is on the move toward nanoscale. By shrinking components to below 1/10,000th the width of a human hair, manufacturers could make faster computer chips with more fire-power per square inch. However, the technology for making devices in that size range is still being developed, and the smaller the components get, the more expensive they are to produce.

Purdue chemist Alexander Wei may have come up with a surprisingly simple and cheap solution to the shrinking data storage problem. Wei's research team has found a way to create tiny magnetic rings from particles made of cobalt. The rings are much less than 100 nm across—an important threshold for the size-conscious computer industry—and can store magnetic information at room temperature. Best of all, these "nanorings" form all on their own, a process commonly known as self-assembly.

"The cobalt nanoparticles which form the rings are essentially tiny magnets with a north and south pole, just like the magnets you played with as a kid," says Wei, who is an associate professor of chemistry in Purdue's School of Science. "The nanoparticles link up when they are brought close together. Normally you might expect these to form chains, but under the right conditions, the particles will assemble into rings instead."

The magnetic dipoles responsible for nanoring formation also produce a collective magnetic state known as flux closure. There is strong magnetic force, or flux, within the rings themselves, stemming from the magnetic poles each particle possesses. But after the particles form rings, the net magnetic effect is zero outside. Tripp developed conditions leading to the self-assembly of the cobalt nanorings, then initiated a collaboration with Dunin-Borkowski to study their magnetic properties. By using a technique known as electron holography, the researchers were able to observe directly the flux-closure states, which are stable at room temperature.

"Magnetic rings are currently being considered as memory elements in devices for long-term data storage and magnetic random-access memory," Wei says. "The rings contain a magnetic field, or flux, which can flow in one of two directions, clockwise or counterclockwise. Magnetic rings can thus store binary information, and, unlike most magnets, the rings keep the flux to themselves. This minimizes crosstalk and reduces error during data processing."

When you turn on your computer, it loads its operating system and whatever documents you are working on into its RAM, or random-access memory. RAM is fast, enabling your computer to make quick changes to whatever is stored there, but its chief drawback is its volatility—it cannot perform without a continuous supply of electricity. Many people have experienced the frustration of losing an unsaved document when their computer suddenly crashes or loses power, causing all the data stored in RAM to vanish.

"Nonvolatile memory based on nanorings could in theory be developed," Wei says. "For the moment, the nanorings are simply a promising development." Preliminary studies have shown that the nanorings' magnetic states can be switched by applying a magnetic field, which could be used to switch a nanoring "bit" back and forth between 1 and 0. But according to Wei, perhaps the greatest potential for his group's findings lay in the possibility of combining nanorings with other nanoscale structures.

"Integrating the cobalt nanorings with electrically conductive nanowires, which can produce highly localized magnetic fields for switching flux closure states, is highly appealing." he says. "Such integration may be possible by virtue of self-assembly."

Several research groups have created magnetic rings before but have relied on a "top-down" manufacturing approach, which imposes serious limitations on size reduction. "The fact that cobalt nanoparticles can spontaneously assemble into rings with stable magnetic properties at room temperature is really remarkable," Wei says. "While this discovery will not make nonvolatile computer memory available tomorrow, it could be an important step towards its eventual development. Systems like this could be what the data storage industry is looking for."

Wei's group is associated with the Birck Nanotechnology Center, which will be one of the largest university facilities in the nation dedicated to nanotechnology research when construction is completed in 2005. Nearly 100 groups associated with the center are pursuing research topics such as nanometer-sized machines, advanced materials for nanoelectronics, and nanoscale biosensors.

2.15   MORE EFFICIENT BASE STATIONS

As mobile devices get better, researchers are also looking to improve the technology that handles users' calls. For example, Cambridge, Massachusetts-based Vanu has created the Vanu Software Radio, a software-based system that promises to replace a mobile phone tower's room full of communications hardware with a single computer. The system is designed to making personal communications more affordable, particularly for small, rural communities. The software is also capable of running emergency communications—such as police, fire, and ambulance channels—on the same device as the civilian system, eliminating the need for a separate network of emergency communications towers. "Rural customers are the first application of the technology, but large carriers are watching to see what happens," says John Chapin, chief technology officer at Vanu.

Vanu scientists developed and tested the software with funds from the National Science Foundation, the federal agency that supports science and engineering research and education. Although not yet commercially available, the technology is beginning to attract the attention of service providers nationwide. "When the telecom industry crashed, Vanu technology caused wireless operators to look at deployments differently," says Sarah Nerlove, the NSF Small Business Innovation Research program officer who oversees Vanu's awards. "Vanu was an ideal fit for their changing needs."

Mobile phone towers dot the landscapes of cities and suburbs, providing millions of Americans with access to wireless communications. At the base of each tower is an air-conditioned shelter filled with expensive equipment called a base station. "As technology advances, all of that equipment continually needs to be overhauled or replaced," says Chapin. Besides replacing much of a base station's hardware with a single server, radio software can aggregate equipment from many stations into a single location that communications engineers call a "base station hotel."

Vanu Software Radio performs all of the functions of a global system for mobile communications (GSM) base station using only software and a non-specialized computer server. The servers run the Linux operating system on Pentium processors, further simplifying the technology and reducing cost. Vanu is demonstrating the technology in two rural Texas communities: De Leon in Comanche County and Gorman in Eastland County. When the test ends, sometime in 2004, the technology will remain as a cellular infrastructure run by Mid-Tex Cellular.

Although the software currently runs on large servers, the product can also be used on a variety of ordinary desktop computers. This attribute will allow service providers to install the software on low-priced systems. Even an off-the-shelf PC can run the software, notes Chapin, although it wouldn't be able to handle a large number of customers. The software's portable design also allows it to easily adapt to hardware upgrades.

The software has carried phone calls since it was installed in the Texas towns in June 2003. Vanu's researchers are now tracking how many calls are successfully handled through the system, how well mobile phones can communicate with other mobile phones, and how well mobile phones can communicate with landline phones.

In the years ahead, large carriers could use the software to establish base station hotels or to upgrade and condense their existing equipment. Additionally, the technology will allow service providers to more efficiently use their portion of the radio frequency spectrum and to quickly adjust to frequency and bandwidth modifications.

2.15.1 Boosting Mobile Phone Range

A new base station remote control system aims to increase the range of mobile phones and also potentially save service operator costs related to operating and repairing defective base station units.

The recent explosive growth in mobile phones has been accompanied by a parallel growth in the underlying networks of base stations used to connect calls. This trend has created headaches for network administrators charged with keeping an increasing number of base stations active at all times. Now, a new power and management device is designed to allow administrators to manage base station operations remotely, reducing repair times, lowering costs, and improving range.

The system was developed by Amper Soluciones, a Spanish company with expertise in telecom network management systems, and Ascom Energy Systems, a German company that specializes in industrial power plants. "Base stations for mobile phone networks are normally located in places where access is quite difficult," says Juan Carlos Galilea, Amper Soluciones' technical and technological support director. "With our system, the operator can remotely determine the real problem in the base station and monitor other systems, such as alarms and communication lines, as well as air conditioning, an external beacon, and even whether the door is open." Some of the detected problems can be solved remotely, whereas others can be solved by maintenance staff on site.

The control unit is built into a small cabinet and offers at least 25 percent more power in the same volume than existing units, says Galilea. The extra power increases the range of the base station, and the small size means that the station can be installed in awkward locations, such as gas stations or church spires. A battery subsystem can maintain operation even with a power loss.

The unit's remote management strengths show through in daily station maintenance, says Galilea. He notes that administrators can monitor their base stations continually and fix any problems as they arise.

Galilea stresses the importance of using software simulations to speed up the design process. Rather than build complete prototypes, the project partners used computer simulations to adjust the density of elements in the power system and keep the operating temperature under control. "Simulations and then mechanical prototypes were used to determine the final structure. This allowed us to reduce development costs," he says. The partners now aim to supply the unit to network operators in Europe and around the world.

Chapter *3*

Connections in the Air— Wireless Technologies

The mobile revolution is being propelled forward by the simultaneous evolution of a set of key technologies in areas such as phone networks, wireless local area networks (WLANs), personal-area networks (PANs), and software infrastructure. Gartner, a technology research firm based in Stamford, Connecticut, reports that core technologies are evolving quickly with little prospect of significant stability before 2005. New developments in areas such as screens, fuel cells, and software for tasks such as speech recognition will continue to drive evolution in the long term.

Wireless technology is the primary driving force behind the most powerful and world-altering telecommunications trends. Gartner reports that wireless networking will become ubiquitous with several different technologies and protocols coexisting in the home and office. By 2007, more than 50 percent of enterprises with more than 1,000 employees will make use of at least five wireless networking technologies. "All organizations should develop a strategy to support multiple wireless networking technologies," says Nick Jones, a research vice president for Gartner. "Organizations developing consumer products for mobile networks should look for ways to add value by interacting with other home devices that might become networked, such as televisions, set-top boxes, game consoles, and remote-control light switches."

Telecosmos: The Next Great Telecom Revolution, edited by John Edwards
ISBN 0-471-65533-3   Copyright © 2005 by John Wiley & Sons, Inc.

3.1   WIRELESS LAN "HOTSPOTS"

Today's WLANs represent just the beginning of what will eventually become a wireless world, connecting people to people and people to machines.

Existing wireless "hotspots"—WLANs that allow mobile device users to access the Internet—allow mobile device users to Web surf, check their e-mail, and swap files while in public places like stores or airports. By 2025, separate hotspots will merge into a "hotworld," enabling people to access the Internet from just about any location on the planet. "People will come to expect continuous connectivity in the way they currently expect to find electric lights wherever they travel," says Martin Weiss, chairman of information science and telecommunications at the University of Pittsburgh.

The past few years have been an extraordinary period for the hotspot market. Hotspots offer an inexpensive way for service providers to drive subscriptions for an increasingly mobile but data-reliant workforce. The number of worldwide hotspots grew from under 2,000 locations to over 12,000 locations in 2002, according to the Scottsdale, Arizona-based market research company In-Stat/MDR. In most regions, hotspot deployment growth continued strong throughout 2003.

Much of the hotspot growth in 2003 resulted from carriers and other large players entering the market. Several European service providers are expected to become more active in the hotspot market in 2003, and providers in the Asia Pacific region will continue to demonstrate a high level of interest. The North American market will be largely impacted by the realization of Project Rainbow. Project Rainbow, a nationwide hotspot network, is supported by AT&T, IBM, and Intel-backed Cometa Networks.

The arrival of 802.11.b "Wi-Fi" wireless has given today's PC users a small taste of what a true "smart home" will be like. Tomorrow's home networks will go beyond file and Internet access sharing to provide wall-to-wall control over home entertainment, information, communications, and environmental and security systems. "We are all going to have a home server, just like the furnace in the basement," predicts Brian Costello, president of Supernova, an Internet consulting company located in Addison, Illinois. "Our computing devices will be tied into that server, along with our refrigerator, microwave, and heating, cooling, and security systems.

3.2   WLANS TO COME

Beyond today's Wi-Fi 80211.b technology, additional 80211.x standards promise to make wireless communication faster and more robust and efficient; these are important considerations for enterprises that are increasingly finding their present wireless LANs strained to the breaking point. Already available, 802.11a supports data rates of up to 54 Mbps. Widespread use, however, has been hampered by incompatibility with 802.11b technology (the standards use different frequency ranges); thus a new standard was developed: 802.11g. This technology provides 802.11a-level data rates along with full 802.11b backward compatibility. The first 802.11g products started appearing in 2003, and the market is expected to shift into top gear by 2005. As prices begin falling, the new standard is expected to gradually edge out 802.11b technology.

In the near future, support is likely to begin appearing for 802.11f, a standard that provides interoperability between access points manufactured by various vendors, enabling portable device users to roam seamlessly between networks. And the alphabet soup doesn't stop there. An array of additional 802.11x standards, covering everything from quality of service (802.11e) to security (802.11i) to network performance and management (80211.k), are also expected to enter the mainstream over the next 12 to 36 months.

Also on the horizon is 802.16. The WiMax standard enables wireless networks to extend as far as 30 miles and transfer data, voice, and video at faster speeds than cable or DSL. It's perfect for ISPs that want to expand into sparsely populated areas, where the cost of bringing in DSL or cable wiring is too high.

The future also looks promising for the up and coming low-rate Wireless PAN (WPAN) technology, 802.15.4, and ZigBee. The ZigBee specification, now in development, will define the network, security, and application interface layers, which can be used with an 802.15.4 solution to provide interoperability. ZigBee Alliance members are definitely determined to carve out a piece of the wireless pie for themselves.

According to In-Stat/MDR, quite a bit hinges on the ability of the ZigBee Alliance to deliver a final specification in a timely manner, including completed, successful interoperability tests. If these milestones are not achieved in a reasonable amount of time, other competing wireless technologies could take hold in these markets, such as a yet-to-be-determined low-rate Ultra-Wideband WPAN alternate PHY or a potential Bluetooth "Lite" version. Therefore, there is an impetuous to move forward according to schedule.

According to Joyce Putscher, director of In-Stat/MDR's converging markets and technologies group, "the heightened interest in 802.15.4/ZigBee wireless connectivity could slowly make 'The Jetsons' home of the future a reality; however, I doubt we'll see that automated meal maker any time soon."

3.3 WLAN FOR EMERGENCY COMMUNICATIONS

Hotspot technology also promises to help public service, emergency services and rescue workers exchange information and collaborate on tasks more effectively and efficiently.

Today, first responders would like to be able to send messages simultaneously to all the emergency workers at the scene of a disaster if necessary, but lack of interoperability among various types of radio equipment prevents them from doing so today. In the future, first responders converging on a disaster scene may be able to quickly and easily exchange emergency messages and data using a wireless ad hoc network recently developed and tested by scientists and engineers at the National Institute of Standards and Technology (NIST). NIST's work in this area is part of the federal government's efforts to improve first responder communications in light of the September 11 terrorist attacks.

The network consists of personal digital assistants (PDAs) equipped with WLAN cards. Transmission routes among the PDAs are established automatically and without need for networking infrastructure at the emergency site as the first responders arrive on the scene. The network may use any nearby PDA to relay messages to others at the scene and allows transmission of voice, text, video, and sensor data.

If a worker leaves the disaster scene or a device is destroyed, the network automatically reorganizes itself. Small video screens can display the names of workers and their roles. In buildings equipped with radios at reference locations, the network would determine the locations of first responders and track their movements. The devices could also receive information from smoke, heat, or vibration sensors embedded in smart buildings that could be transmitted by wireless sensor networks or distributed by first responders during emergencies.

3.4 SMART BRICK

Wireless technology can also be used to prevent an emergency before it happens. A "smart brick" developed by scientists at the University of Illinois at Urbana-Champaign, for example, monitors a building's health and uses a wireless link to relay critical information that could save lives.

In work performed at the school's Center for Nanoscale Science and Technology, Chang Liu, a professor of electrical and computer engineering, and graduate student Jon Engel have combined sensor fusion, signal processing, wireless technology, and basic construction material into a multi-modal sensor package that can report building conditions to a remote operator.

The innovation could change the face of the construction industry, says Liu. "We are living with more and more smart electronics all around us, but we still live and work in fairly dumb buildings. By making our buildings smarter, we can improve both our comfort and safety."

The prototype smart brick features a thermistor, two-axis accelerometer, multiplexer, transmitter, antenna, and battery. Built into a wall, the brick could monitor a building's temperature, vibration, and movement. Such information could be vital to firefighters battling a blazing skyscraper or to rescue workers ascertaining the soundness of an earthquake-damaged structure.

"Our proof-of-concept brick is just one example of where you can have the sensor, signal processor, wireless communication link, and battery packaged in one compact unit," says Liu. "You also could embed the sensor circuitry in concrete blocks, laminated beams, structural steel, and many other building materials."

To extend battery life, the brick could transmit building conditions at regular intervals instead of operating continuously. The brick's battery could also be charged through the brick by an inductive coil, similar to the type used in electric toothbrushes and some artificial heart pumps.

The researchers are currently using off-the-shelf components, so there's plenty of potential for making a smaller sensor package. "Ultimately, we would like to fit everything onto one chip, and then put that chip on a piece of plastic, instead of silicon, to make it more robust," says Engel. Silicon is a rigid, brittle material, which can easily crack or break. Sensor packages built on flexible substrates would not only be more resilient, they would offer additional versatility. "For example, you could wrap a flexible sensor around the iron reinforcing bars that strengthen concrete and then monitor the strain," says Engel. The researchers have already crafted such sensors by depositing metal films on flexible polymer substrates. Besides keeping tabs on a building's health, potential smart brick applications include various other types of monitoring chores. "In the gaming industry, wireless sensors attached to a person's arms and legs could replace the conventional joystick and allow a 'couch potato' to get some physical exercise while playing video games such as basketball or tennis," says Liu. "The opportunities seem endless."

3.5 WIRELESS SMART STUFF

Wireless networks, along with improvements in processor, software, and related technologies, will lead to the arrival of smart appliances. A smart appliance is a household device that supplements its basic function, such as keeping food cold, with internal intelligence and external communication capabilities.

A smart refrigerator, for example, could keep track of product quantities and expiration dates, via a code embedded into the product packaging, to make sure that there's always an ample and fresh supply of food and beverages. Using a wireless link to the home's central server, a smart refrigerator could automatically notify its users of products that must be purchased on the next shopping trip. The refrigerator could even place orders directly to merchants for home delivery. By 2025, smart refrigerators, as well as smart toasters, microwave ovens, washing machines, clothes dryers, and dish washers, should all be commonplace.

Wireless broadband links between TVs, stereos, household appliances, and other devices will allow information and entertainment systems to share data, making them all highly interoperable. Forget about searching for a favorite song or movie—the files will be stored inside the home's central server. The server will also allow users to manage all connected devices through table-top and wall-mounted displays, as well as portable devices.

We can also expect the home of tomorrow to be equipped with cameras, perhaps in each room. "You're going to see a lot more people with video cameras in their home," says Jennifer Sterns, a marketing manager at TDS Metrocom, a communication services provider located in Madison, Wisconsin. "They can watch from work so they can see their babbysitters."

3.6 WIRELESS ON WHEELS

Engineers at the University of California-San Diego (UCSD) have created the world's first bus that allows its passengers to access the Internet and download files at a peak speed of 2.4 Megabits per second—even while the vehicle is moving.

The broadband wireless bus dubbed the "CyberShuttle" combines a fully mobile 802.11b wireless local area network inside the bus, with Web access through QUALCOMM's CDMA2000 1xEV wireless wide-area data network installed on the UCSD campus and at the company's San Diego headquarters. "Our students and faculty are getting a taste today of wireless technology that most of the world will not be using until years from now," says Elazar Harel, the university's assistant vice chancellor for administrative computing and telecommunications. "This bus is one of the first places where we will be able to experiment with technical as well as social aspects of third-generation [3G] wireless services in a real-world environment."

The commuter bus shuttles students, faculty, and visitors between the main UCSD campus in La Jolla and the Sorrento Valley train station. The trip typically takes 15 to 20 minutes, enough time for commuters to check e-mail or surf the Web. Passengers can also watch streaming video or listen to high-fidelity music because they are connected to the Internet in a dedicated 1.25-MHz channel at speeds up to 2.4 Megabits per second. "The campus already has 1,200 users registered to use 802.11, and all of them can just as easily log on while riding the bus," says Greg Hidley, head of technology infrastructure. "To use their laptops or personal digital assistants online, all they need is the same wireless card they use elsewhere on campus, and they are automatically handed off from the local network, through the 1xEV network, to the Internet."

3.7 MESH NETWORKS

A mesh network is a network in which there are at least two pathways to each node. A fully meshed network means that every node has a direct connection to every other node, which is a very elaborate and expensive architecture. Most mesh networks are partially meshed and require traversing nodes to go from

Figure 3-1   Mini Talker.

each one to every other. Mesh networks provide redundancy by supplying multiple pathways to each node.

Mesh networks are very well suited for use in vehicle monitoring systems, such as the one developed by MachineTalker. The Goleta, California-based start-up has developed MiniTalker, a series of mesh-network-based modular wireless devices that automatically create ad hoc, peer-to-peer WLANs inside vehicles (Fig. 3-1). The devices, which feature built-in environmental sensors, are designed for use on a variety of vehicles—including aircraft, cars, trucks, and trains—and in shipping containers, to measure and report on conditions such as temperature, humidity, and vibration levels. The MiniTalkers will come in various forms for different applications, including stand-alone units that ride along in a vehicle, collecting data, and "tags"—dubbed TagTalkers—that can be attached to a vehicle or a component with Velcro.

MiniTalkers are based on the company's MachineTalker, a mesh network concept. MachineTalker's technology enhances the mesh network structure by eliminating the need for a powerful, power-hungry CPU. A microcontroller inside each MiniTalker allows the formation of a "local community," with each unit having the ability to freely share information and spot defective network nodes. The proprietary Simple Machine Management Protocol (SMMP) handles network functions. A central unit gathers the WLAN's data and links the information to an external network, such as the Internet or an intranet.

MiniTalker is designed to give airlines and trucking companies the ability to continuously monitor key vehicle components, such as pumps and gear assemblies, as well as cargo. Roland Bryan, MachineTalker's president and CEO, describes the units as "intelligent proxies." "They talk for whatever they're attached to," he says. Bryan says MiniTalker prototypes have been demonstrated to several vendors, although the company has yet to snag any customers.

MachineTalker's initial strategy is to target airlines, which can use the technology to gather airframe vibration data, from takeoff to landing. The company also plans to provide MiniTalker-based security services to companies that own and transport shipping containers. Bryan also believes that the technology has potential stationary applications, such as networking a building's vending machines, allowing continuous remote monitoring of product temperature, availability, and other conditions.

A mesh WLAN marks an improvement over radio frequency identification (RFID) systems, which can track mobile assets but lack the capability to self-diagnose problems or sense real-world environmental conditions. Additionally, on some types of vehicles, particularly aircraft, mesh WLANs can be less expensive to install and maintain than wired LANs. "If you've ever looked under the skin of an airplane, there are wires everywhere," says Thomas Turney, an investment banker with NewCap partners, a Los Angeles venture capital firm. "It's extremely expensive to add wired components." MachineTalker faces several components, most notably MeshNetworks and PsiNaptic, which also offer or plan to offer wireless networks with monitoring capabilities.

MiniTalker incorporates several components. Each device includes a microprocessor, a miniature radio, and sensors. The MiniTalker prototype uses an ATmega 128L, an 8-bit RISC microcontroller with 128 kilobytes of flash memory, from San Jose-based Atmel. The radio is a single-chip 900-MHz transceiver from Dallas-based Texas Instruments Inc., although MachineTalker is looking to switch to a Xemics XE1202 single-chip transceiver for its production models. "It has higher power output and a more sensitive receiver," says Bryan.

A pair of companies supplies MiniTalker's sensors. The temperature/humidity sensor comes from EME Systems of Berkeley, California. Norwood, Massachusetts-based Analog Devices provides an accelerometer for vibration sensing. Production models may include other sensors from other vendors. "We're looking at little, tiny stuff," says Bryan. Turney predicts that MiniTalker and other future mesh WLANs will create a growing market for sensors that can measure sound, light, gas, liquid, and various other levels. "There will be a large demand for sensors," he says.

Will MiniTalker fly in the market as well as on aircrafts? Turney foresees "almost limitless" growth for the WLAN market and its suppliers. He says he's not sure that mesh WLANs will ever be as cheap as RFID. "But for lots of applications for which you need more intelligence, I see this as a big opportunity," he says.

Thilo Koslowski, lead automotive analyst for GartnerG2, the business strategy division of Gartner, also believes that mesh WLANs have strong potential. "We will have devices, products, technology that will talk to each other." But he notes that it's not easy for a single start-up to jump-start an entire market. "It's very difficult for one small company to really push for that," he says.

3.7.1 Emergency Mesh

Mesh networks can also be used to protect people. Toxic clouds, whether caused by an industrial accident, transportation mishap, or terrorist attack, can threaten any metropolitan area. But a new telecom-driven defense technology, currently being planned by researchers at Cornell University, promises to help local authorities determine the extent and direction of a chemical- or biological-based outbreak.

With this technology, helicopters are into affected areas and then release flurries of tiny devices—each about the size of a dime. Analysts will be able to determine with these devises an attack's scope, uncover hazardous hotspots, and track the direction of toxic gasses or biohazards. The devices contain sensors that sample the air for toxins as well as tiny radio transceivers that allow the devices to communicate with each other. By relaying data, the devices report to a human operator at the fringe of the disaster area; a display shows where the contamination is and how it's spreading.

The research project brings together molecular biologists, device physicists, telecom engineers, information and game theorists, and civil engineers to develop self-configuring sensor networks for disaster recovery. This initial research will focuses on the detection of biohazards; however, the underlying principles can also be applied to many other critical situations, including searches for earthquake victims (using audio and body-heat sensors) and the monitoring of municipal water systems for leaks or contamination.

In the aftermath of a disaster, the most pressing need is for information. Because it is often dangerous or even impossible to collect data manually in a disaster, the researchers' plan is to create an automated self-configuring remote sensor network. The idea grew out of studies of how such networks could be used on the battlefield, but the project focuses on civilian applications. "If they can save the lives of soldiers, you can use them to save the lives of civilians," says Stephen Wicker, the Cornell professor of electrical and computer engineering who heads the research team.

Two types of biosensors are being developed, capable of detecting a variety of agents, including toxins and bacteria, using biological material incorporated into silicon microcircuits. One type uses a membrane topped with binding sites, like those on the surface of a living cell, layered onto a silicon microcircuit. When a molecule, such as a neurotoxin, binds to the surface—just as it would when attacking a living target—a protein channel is opened, allowing ions to

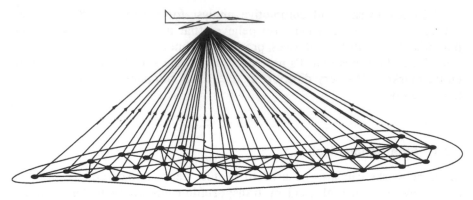

Figure 3-2 *Sensors use the strength and direction of radio signals from their neighbors to map their locations.*

flow through the membrane and creating an electrical signal detected by the underlying circuit. The other sensor under development contains antibodies placed between tiny electric contacts. When an agent such as a virus, bacteria, or spore binds to the antibody, the current flowing between the contacts is altered.

To keep size and power requirements small, the devices will communicate using very-low-power radio signals, allowing each device to reach only a few others in its immediate neighborhood. Rather than using global positioning system (GPS) technology, which would add weight and power requirements, the sensors will use the strength and direction of radio signals from their neighbors to map their locations. The signals will be relayed from one device to another until they reach a human operator at the edge of the territory. In theory, the "reachout point" could be a van, a low-flying aircraft, or even a satellite (Fig. 3-2).

The engineers will also draw on game theory—which deals with how a group of individuals interacts and competes for resources—to program the devices to work together. The devices will decide, for example, the order and direction in which messages should be relayed, avoiding redundant signals. If the coverage mapping shows that some areas are not covered, the network will be able to call for the deployment of additional sensors.

Additionally, the researchers will draw on case histories of earthquake effects, accidents in crowded urban environments, and the aftermath of the World Trade Center attack to develop prototype applications, which in turn will determine the design of the sensors and networks. This work will draw on a database developed by the Multidisciplinary Center for Earthquake Engineering Research at the University of Buffalo as well as on interviews with experienced rescue workers, including some who worked in the wreckage of the World Trade Center.

3.8   WIRELESS SENSOR IS A "SPEC"

Ultimately, a mesh network is only as good as the sensors it incorporates. Small, lower-power sensors are the keys to making a mesh network more useful and productive. Adhering to this principle, University of California (UC) at Berkeley researchers have developed a wireless sensor that can fit on a fingertip.

The device, dubbed "Spec," is a "mote"—a wireless sensor platform that measures a mere $5\,mm^2$. Massive numbers of these tiny devices could be used in self-organizing wireless sensor networks for such innovative applications as monitoring seabird nests in remote habitats, pinpointing structural weaknesses in a building after an earthquake, or alerting emergency workers to the presence of biochemical toxins.

"Spec is our first mote to integrate radio frequency communication and custom circuits onto a chip that runs the TinyOS operating system," says Kris Pister, a UC Berkeley professor of electrical engineering and computer sciences and the researcher heading the project. "It's a major step towards sensors that cost less than a dollar apiece and that are integrated into the products that we own, the buildings that we live and work in, and the freeways we drive on. The potential for such sensor networks is enormous."

What's remarkable about Spec is the range of components the researchers were able to fit onto a single chip and its system-driven architecture. Spec combines a micro-radio, an analog-to-digital converter, a temperature sensor, and the TinyOS operating system onto a piece of silicon 2 by $2.5\,mm^2$. "This is the first fully integrated and fully operational mote on an individual chip," says David Culler, a UC Berkeley computer science professor. "Single chip integration makes the mote very cheap because it reduces post-assembly requirements. This opens the path to very low cost deployments of a large number of motes."

Researchers tested the new chip at the Intel Research Laboratory in Berkeley. Spec was able to transmit radio signals at 902 MHz over 40 feet. "We were in a lab environment with a lot of high-power equipment that generates interference," says researcher Jason Hill, who received his PhD in electrical engineering and computer sciences from UC Berkeley in May. "If we went outdoors and had direct line of sight between the two motes, we could realistically transmit about twice the distance we did indoors."

Spec also includes hardware accelerators to help the core pieces of TinyOS run more efficiently. The accelerators allow data encryption to be performed by the hardware, which is thousands of times more efficient than performing the same function in software. The mote's radio transmitter uses 1,000 times less power than a standard mobile phone.

In addition to the chip, Spec requires an inductor, an antenna, a 32-kHz watch crystal, and a power source. The researchers note that these components would add little to the mote's size. Several companies are already developing millimeter-scale batteries that could power a commercial version of Spec.

Besides its academic and commercial potential, Spec has generated interest from the military for possible uses on the battlefield or to monitor troop movements. Pister says he plans to develop Spec into a commercial product within the next year. He notes that the finished product, which will include the battery and casing, will likely be about the size of an aspirin.

3.9   COLLABORATIVE SENSING

Large-scale, distributed and sensor-rich wireless networks can be used to track physical phenomenon, including multiple moving objects such as vehicles or animals. Potential applications include traffic control, battlefield target tracking, security, and monitoring of wildlife, environmental pollutants, and infrastructures such as power and telecom grids.

Scientists in the embedded collaborative computation group of Xerox's Palo Alto Research Center (PARC) are exploring problems in information processing, communication, storage, and routing in environmentally based smart sensor networks. Such networks are constrained by energy and bandwidth limitations, which require solutions to take resource conservation into account. Because sensors may be exposed to the wear and tear of motion and changing weather, there are also issues of robustness.

Decentralized systems require the identification of structures within collections of spatio-temporally distributed signal streams. A critical technical challenge is to aggregate, represent, and maintain the structure-level information from point-wise sensor data in an irregular, dynamic network. This requires approaches beyond traditional signal processing.

PARC researchers are developing novel data interpretation approaches that blend statistical and model-based techniques for explicitly reasoning about uncertainty and resources. They are employing artificial intelligence to enable individual sensors to "know" what information to sense and share.

PARC's collaborative processing approach is inherently scalable. One aim of the research is to develop the capability to aggregate, store, and process information from hundreds of thousands of sensors. PARC researchers are engaged in talks with customers about real-world implementations of their collaborative sensing technologies. One potential application would embed sensors in roads and vehicles to manage traffic flow. Another would create smart security systems that could detect intrusions in buildings or within the boundaries of airports, power plants, or other sites. Smart sensors could also be used to track the movement of supplies and equipment.

3.10   OPTICAL SENSORS

As more sensors are built into various types of networks, developers and users are beginning look for devices that can collect the most data while consuming the least power and occupying the smallest possible space. Fiber-optic sensors fulfill all of these criteria and more. Fiber-optic sensors offer a wide spectrum of advantages over traditional sensing systems, including smaller sizes and longer lifetimes. Immunity to electromagnetic interference, amenability to multiplexing, and high sensitivity make fiber optics the sensor technology of choice in several fields, including the health care and aerospace sectors.

Optical systems require physically smaller media for representing information than is required by magnetic or electronic systems. This requirement gives them an edge over conventional devices. The greater bandwidth of optics enables delivery of more data, which is useful for high-speed data transmission or high-resolution video transmission.

"Optical sensors are not only replacing conventional sensors in many areas in science, engineering, and medicine but researchers are also creating new kinds of sensors that have unique properties," says Joe Constance, an analyst at Technical Insights, a technology research firm located in San Antonio, Texas. "These properties relate to the ability of the sensors to measure physical, chemical, and biological phenomena."

Electromagnetic interference can corrupt data transmitted from a conventional thermocouple. Fiber-optic sensors, on the other hand, show greater resistance than thermocouples to hostile environments and electromagnetic interference, making them an ideal choice as temperature sensors in many applications. Scientists have been working on a fiber-optic sensor that measures temperature with a reflector for use in industrial power plants, nuclear plant, aircrafts, and ships. "Researchers are intent on further improving the bond between the fiber and the reflector, as well as reducing the required electronics for data acquisition and analysis," says Constance.

Recent advances in fiber optics and the numerous advantages of light over electronic systems have boosted the utility and demand for optical sensors in an array of industries. Environmental and atmospheric monitoring, earth and space sciences, industrial chemical processing and biotechnology, law enforcement, digital imaging, scanning, and printing are only some of them. The ubiquity of photonic technologies could drive down prices as they have done in the telecommunications market, which reduced the cost of optical fibers and lasers.

3.11 NAVIGATING THE REAL WORLD

University of Rochester researchers have created a navigational assistant that can help inform a visually impaired person of his or her whereabouts. The technology could also be used to add new dimensions to museum navigation or campus tours for sighted people.

The Navigational Assistance for the Visually Impaired (NAVI) system uses radio signals to gauge when someone is near a passive transponder, which can be as small as a grain of rice. The transponder might be located on the outside of a building, inside a hallway, or on a painting or other object of interest. The system works much like a retail security tag or the cards people wave at specially equipped gas pumps to make fast, automatic purchases. With such systems, a radio signal, beamed from a detector located near a door or gas pump, is picked up and returned by the tag. The security tag system, when activated, simply sets off an alarm. Purchase tags, on the other hand, can encode information, allowing the reader to debit the user's bank or credit card account to complete the sale.

With the NAVI system, researchers have decided to turn things around and have made the reader portable and affixed tags to stationary objects. The prototype reader uses a portable CD player that's programmed to play a particular track through an earphone whenever a certain tag is detected. A particular track might contain a simple message, such as, "Mr. Smith's office door," an elaborate discussion concerning a particular piece of art, or the history of an entire building. Users can switch CDs for different purposes and locations. Production versions of the device could store information in solid-state memory—like the type used in MP3 players—that's instantly updated via a wireless link whenever the user enters a new building or exhibit area.

Built with off-the-shelf components, the prototype NAVI device is a black box about half the size of a loaf of bread. The reader includes the CD player plus an antenna that looks somewhat like a singer's microphone. A final version would probably be about as small as a portable CD player. If solid-state memory were incorporated, the entire device could be no larger than a deck of cards.

"To prepare a building or site for use with this system will be relatively inexpensive," says Jack Mottley, a University of Rochester associate professor of electrical and computer engineering. "The tags are inexpensive now and the prices are still dropping. The plan is to use only passive tags that do not require batteries or need to be plugged in, meaning once they are installed they can be ignored." Tags could even be painted over without losing their capabilities.

Down the road, NAVI may find additional uses. The technology could, for example, be built into mobile phones or wristwatches. This would allow users to gain information on almost anything around them, from customer reviews about a shirt they're considering buying to automatically paying for a soda at a vending machine.

3.12 WIRELESS UNDERWEAR

In an effort designed to protect the health of chronically-ill people, personal garments—even underwear—could someday be equipped with wireless technology.

Scientists at Philips Research in Aachen, Germany, have developed a wearable, wireless monitoring system that can alert patients with underlying health problems to a potentially serious situation, assist clinicians in the diagnosis and monitoring of "at-risk" patients, and automatically contact emergency services in the event of an acute medical event. Based on dry-electrode technology that can be built into common clothing items, including bras, briefs, or waist belts, Philips' wireless technology continuously monitors the wearer's body signals, such as heart activity, to detect abnormal health conditions. The technology could pave the way to the development of a new category of products in the personal healthcare area.

When worn continuously by the patient, the wireless monitoring system can store up to three months of body signal data, such as heart-rate information, in 64 MB of internal memory, thus providing clinicians with a continuous history over an extended period of time to assist in an accurate diagnosis. As the system is worn, advanced analysis algorithms, executed on the system's ultra-low power consumption digital signal processor (DSP), continuously monitor and record any abnormal signal. If a potentially serious health condition is detected, the system can trigger local alarms or wirelessly link with mobile phone or public switched telephone networks to summon immediate help.

All the system's active electronics are incorporated into an ultra-slim module that slips into a dedicated pocket sewn into the garment. After the module has been removed, the garment—still containing the built-in dry electrodes—can be laundered. Dry electrode technology was designed as a practical, long-term-wearable, alternative to conventional gel-based electrocardiogram pickups. The technology was originally developed by NASA scientists in the early 1960s to monitor astronaut heart activity on multi-day missions. Due to the dry-electrodes' difficulty to harness ultra-low voltages and several other tricky design issues, it was not until the mid-1990s that the technology became commercially viable.

Philips claims that the new system fits well into its vision of "Ambient Intelligence." The company's strategy is based on technologies that disappear into the fabric of their surroundings to improve the quality of users' lives—in this case, personal health-care.

Chapter *4*

The Future is Fiber—
Optical Technologies

Photonics is increasingly being identified as a key enabling technology in the construction of a global communications network. Its potential in advancing information systems and image-processing technologies is expected to stimulate photonics-based research and development initiatives.

Besides generating and controlling light, photonics technology can rapidly disseminate and process large volumes of digital information. Speed, immunity from interference, increased bandwidth, and enhanced data storage capacity are some of the advantages of working with light. These attributes are channeling sizeable investments into photonic research activities. "A single optical fiber can carry the equivalent of 300,000 telephone calls at the same time," says Michael Valenti, an analyst with Technical Insights, a technology research firm based in San Antonio, Texas. "Photonics technology also provides sufficient communication capacity to meet the forecast demand for fully interactive multimedia, Internet services." The rapid transition from an electronic to an optical telecommunications network is anticipated to spur multidisciplinary efforts to take advantage of the advanced information-carrying ability of photons.

4.1 FASTER NETWORKS

New research in laboratories on opposite U.S. coasts shows that optical telecom technology is still far from reaching its full potential. At the Massa-

Telecosmos: The Next Great Telecom Revolution, edited by John Edwards
ISBN 0-471-65533-3   Copyright © 2005 by John Wiley & Sons, Inc.

chusetts Institute of Technology, a research team led by physicist Evan J. Reed has discovered that sending a shockwave through a photonic crystal allows far greater control over light. When the light between a moving shock front and a reflecting surface is confined, incoming light can become trapped at the shockwave boundary, bouncing back and forth in a "hall of mirrors" effect. As the shock moves through the crystal, the light's wavelength is shifted slightly each time it bounces. If the shockwave travels in the opposite direction of the light, the light's frequency will get higher. If the wave travels in the same direction, the light's frequency drops.

A photonic crystal is a credit-card-thick stack of optical filters. By changing the way the crystal is constructed, a user could control exactly which frequencies go into the crystal and which come out. The shockwave approach could be used to efficiently convert light into a wide range of frequencies useful for communications purposes, offering the potential for faster and cheaper telecom devices.

Meanwhile, at the University of California at Santa Barbara (UCSB), researchers have for the first time incorporated both a widely tunable laser and an all-optical wavelength converter onto a single chip, creating an integrated photonic circuit for transcribing data from one color of light to another. Such a device could prove to be the key to creating an all-optical network.

Optical fibers transport information between cities via optical fibers. Each fiber can move numerous colors of light simultaneously, with every color representing a "dedicated" transmission line. As data moves between coasts through Internet nodes, the need exists to switch colors. Information arriving on one fiber as orange photons, for example, may need to continue the next leg of its journey on another fiber as red photons because the channel for orange on that fiber is in use. Switching from one color to another is currently accomplished by converting photons into electrons, making the switch electronically, and then converting the electrons back into photons. The new postage-stamp-size device developed by a research team led by Daniel Blumenthal, a UCSB professor of electrical and computer engineering, is a tunable "photon copier" that eliminates electronics as the middleman.

Past attempts at engineering photonic circuits with tunable lasers and wavelength converters have had only limited success, and the two components currently exist on separate chips. Integrating and fabricating these circuits on the same indium phosphide platform could eventually lead to lower equipment costs and the ability to avoid signal degradation that may occur when light is moved between chips.

4.1.1 Faster Fiber

A team of Florida Tech researchers has developed a technique that can quadruple the amount of information that can be carried on single fiber optic cable.

With the use of a process called spatial domain multiplexing, designed by Syed Murshid, an associate professor of engineering, Barry Grossman, a professor of electrical engineering, and doctoral graduate assistant Puntada Narakorn, one fiber optic cable can transmit multiple pieces of information at the same wavelength without interference, thus significantly increasing the effective information-carrying capacity of the cable. "In this process, we are able to transmit information from multiple sources at the same frequency with high reliability and high accuracy," Murshid says. "In effect, we quadruple the information-carrying capacity at a very low cost. The technology could be the biggest fiber optic communication breakthrough since the development of dense wavelength division multiplexing (DWDM), a technique that uses multiple lasers to transmit several wavelengths of light simultaneously over a single optical fiber.

"The information signal carried through fiber optics is a beam of light, much like that projected on the wall," notes Murshid. "In order to prevent a loss of signal over great distances, the glass used must be very clean." The glass is so clean, in fact, that if the ocean was as pure, one could see the bottom from the surface. With spatial domain multiplexing, the information-carrying light pulses are transported through the fiber optic cable as concentric circles, giving the pattern the appearance of a target. "The future of fiber optics is right on target," he notes.

Ron Bailey, dean of Florida Tech's college of engineering, says Murshid's discovery can potentially transform the telecommunications industry. "By increasing the capacity of a single optical fiber, Dr. Murshid's process has eliminated the need for additional cables," says Bailey. "Up until now, if a telecommunications company needed more capacity, it was forced to undergo the expensive process of laying down more fiber. This new technology provides them with a cost-effective solution."

Murshid believes the technique that makes it possible to quadruple the amount of information carried at the same frequency on a single fiber optic cable also has the potential for additional gains in information-carrying capacity. "We've been able to successfully transmit at the same frequency four independent beams of information-carrying light so far," he says. "But we're only scratching the surface. We will be able to increase this number over time."

Murshid compares multiplexed information sent through fiber optics with FM radio. "Radio stations have to broadcast at a certain frequency," he says. "WFIT, for example, is 89.5 MHz here on the Space Coast. But if you go to Tampa, you will hear a different station broadcasting on 89.5 because the distance between the two stations enables them to operate without interference." The same was true for information sent through fiber optic cables, he notes. Each cable could accommodate a set of frequencies or wavelengths but could use each individual frequency or wavelength only once. Now, through spatial domain multiplexing, the same fiber optic cable can transmit multiple pieces of information at the same wavelength without interference.

4.1.2 Next-Generation Telecom Network

Several scientists, working in collaboration, are laying the groundwork for a new wireless and optical fiber-based telecommunications network that aims to bring reliable, high-speed Internet access to every home and small business in the United States within the next few years.

Funded by a five-year, $7.5 million National Science Foundation grant, the "100 Megabits to 100 Million Homes" research project brings together scientists from Carnegie Mellon University, Fraser Research, Rice University, the University of California at Berkeley, Stanford University, Internet2, the Pittsburgh Supercomputing Center, and AT&T Research. The coalition believes that the growing demand for communications, combined with newly emerging technologies, has created a once-a-century opportunity to upgrade the nation's network infrastructure.

"What the copper-wired telephone network was to the 20th century, the fiber network will be to the 21st," says Hui Zhang, the project's principal investigator and a Carnegie Mellon associate professor of computer science. "Today we have 500 kilobits reaching 10 million American homes," he notes. Zhang is looking toward creating a system 100 times faster and that reaches 10 times more households. "We must make the system more manageable, more secure, more economical, and more scalable, and we must create an infrastructure that can support applications not yet envisioned," he says.

The creation of a network to serve 100 million households with two-way symmetric data communications service at 100 megabits per second is a tremendous challenge that reaches far beyond technological issues. Universal availability of such a network promises to bring fundamental changes to daily life and could substantially raise its users' standard of living. Barriers to the network's creation, however, extend beyond straightforward deployment and cost issue—fundamental innovations in the way networks are organized and managed are also an issue. Additionally, the researchers must develop communication architectures that are particularly well suited to very large-scale deployment and that can operate inexpensively at very high speeds.

To achieve their goal, the collaborators plan to start with basic principles and undertake fundamental research that addresses the design of an economical, robust, secure, and scalable 100 × 100 network. Then, they will construct proof-of-concept demonstrations to show how the network can be built. Initially, the scientists will produce a framework of what such a network might look like. Then, the network's proposed architecture and design will be disseminated to government and industry through presentations and partnerships so it can serve as a guide to business investment in network development.

Zhang notes that the physical testbeds created through the project will serve as a basis for further studies, such as social science research on the impact of connectivity in the home. The software and tools used to design and validate the network, particularly the emulation systems, will be used to create new curricula for network education in two- and four-year colleges. In fact, plans are already under way for an outreach program. "We will attack pieces of the problem according to our expertise and get together to hash out the architecture and develop some initial answers," Zhang says. He expects to have an experimental component at the end of the project in five years.

The Internet has made a huge impact on society, but there are limits to today's network technology, notes Zhang. "We must not be satisfied by the Internet's apparent success," he says. "Breakthroughs over the last 30 years have masked its underlying problems. We need to take a fresh look at the architecture considering new requirements and the technology that has changed profoundly in the last three decades. The biggest challenge is to imagine the network beyond the Internet."

4.2 NEW OPTICAL MATERIALS

Organic electro-optic polymers have long held the promise of vastly improving many different types of telecom technologies. It now appears that scientists are on the verge of breakthroughs that will bring dramatic progress in such materials as well as the devices in which they are used.

Electro-optic polymers are used to make devices that take information that has typically been transmitted electronically and transfer it to light-based optical systems. The latest developments will affect not just how much information can be sent at one time but also the power needed to transmit the information.

The capabilities of the most recently developed materials are about five times greater than those of standard lithium niobate crystals, the best naturally occurring material for transferring data from electronic to optical transmission and for many years the industry standard. The newest materials require less than one-fifth the voltage (under 1 volt) needed for lithium niobate. "What this shows is that people have done far better than nature could ever do in this process," says Larry Dalton, a University of Washington chemistry professor and director of the Science & Technology Center on Materials and Devices for Information Technology Research. "The reason we're seeing improved performance is the rational design of new materials with new properties." The newest materials represent a nearly fivefold improvement in capability in just four years. At that rate, material capabilities will soon reach benchmarks set for 2006 by the National Science Foundation.

Recent advancements are making possible technologies that were previously only a fanciful vision, says Dalton. For example, components can now be made so small and power efficient that they can be arranged in flexible, foldable formats yet experience no optical loss or change in power requirements until the material is wrapped around a cylinder as tiny as 1.5 mm, a little bigger than a paper clip.

Such materials can be used to create space-based phased array radar systems for surveillance and telecommunications applications. Each face of a phased array typically has thousands of elements that work in a complex interdependence. A major advantage of the new material is that the entire radar system can be launched in a very compact form and then unfurled to its full form once it reaches orbit. Deployment costs can be greatly reduced because of low power requirements and the much-reduced weight of the material being sent into space. According to Dalton, techniques to mass produce the tiny foldable components, which should reduce costs even further, are currently being developed.

The newest materials have immediate applications in a number of other technologies as well, Dalton says. For instance, photonic elements can make it possible for a mobile phone to transmit a large amount of data with very low power requirements, allowing a device that is very efficient to also be made very compact. Similarly, the materials can bring greater efficiency and affordability to optical gyroscope systems, commonly used in aircraft navigation but also adaptable for other uses—such as vehicle navigation systems—if costs are low enough.

Additionally, photonics can replace coaxial cable in many satellite systems, reducing the weight of certain components as much as 75 percent. "The cost of getting something up into space is horrendous because of weight, so anything that reduces weight and power requirements is of immediate importance," Dalton says.

4.2.1 New Glasses

Researchers have developed a new family of glasses that will bring higher power to smaller lasers and optical devices and provide a less-expensive alternative to many other optical glasses and crystals, like sapphire. Called REAl Glass (rare earth aluminum oxide), the materials are durable, provide a good host for atoms that improve laser performance, and may extend the range of wavelengths that a single laser can currently produce (Fig. 4-1).

With support from the National Science Foundation (NSF), Containerless Research Inc. (CRI), based in the Northwestern University Evanston Research Park in Illinois, recently developed the REAl Glass manufacturing process. NSF is now supporting the company to develop the glasses for applications in power lasers, surgical lasers, optical communications devices, infrared materials, and sensors that may detect explosives and toxins.

"NSF funded the technology at a stage when there were very few companies or venture capitalists that would have made the choice to invest," says Winslow Sargeant, the NSF officer who oversees CRI's Small Business Innovation Research (SBIR) award. "We supported the REAl Glass research because we saw there was innovation there," adds Sargeant. "They are a great company with a good technology, so we provided seed money to establish the

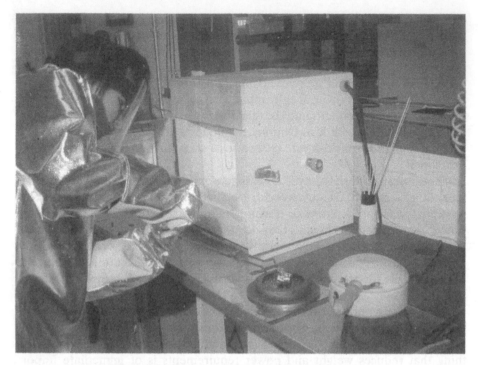

Figure 4-1   REAl Glass (Rare Earth Aluminum oxide).

technology's feasibility. Right now, we can say the feasibility is clear, and they're one step closer to full-scale manufacturability," he says.

CRI originally developed the glasses with funding from NASA. The research used containerless processing techniques, including a specialized research facility—the Electrostatic Levitator—at the NASA Marshall Space Flight Center in Huntsville, Alabama. With the NASA device, the researchers levitated the materials using static electricity and then heated the substances to extremely high temperatures. In that process, the materials were completely protected against contact with a surrounding container or other sources of contamination.

"The research that led to the development of REAl Glass concerned the nature and properties of 'fragile' liquids, substances that are very sensitive to temperature and have a viscosity [or resistance to flow] that can change rapidly when the temperature drops," says Richard Weber, the CRI principal investigator on the project.

REAl Glass, like many other glasses, is made from a supercooled liquid. This means that the liquid cooled quickly enough to prevent its atoms from organizing and forming a crystal structure. At lower temperatures, such as room temperature, the atoms are "fixed" in this jumbled, glassy state. In REAl Glass, the glass-making process also provides a mechanism for incorporating rare-earth elements in a uniform way. This quality makes REAl Glass particularly attractive for laser applications.

After CRI scientists spent several years on fundamental research into fragile liquids, NSF provided funds to develop both patented glasses and proprietary manufacturing processes for combining the glass components in commercial quantities and at a much lower cost than for levitation melting. Using high-temperature melting and forming operations, CRI is making REAl Glass in 10-mm-thick rods and plates, establishing a basis for inexpensive, large-scale production of sheet and rod products.

"The REAl Glass products are a new family of optical materials," says Weber, who adds that CRI is already meeting with businesses to talk about requirements for laser, infrared window, and other optical applications and supplying finished products or licensing the material for use.

"The REAl Glass technology combines properties of competing materials into one [material]," says NSF's Sargeant. "With these glasses," he adds, "researchers can design smaller laser devices, because of the high-power density that can be achieved, and can provide small, high-bandwidth devices for applications in the emerging fiber-to-the-home telecom market."

Because the glass can incorporate a variety of rare-earth elements into its structure, CRI can craft the glasses to yield specific properties, such as the ability to tune a laser across multiple light wavelengths, which can have important implications for the lasers used in dental procedures and surgery, for example, providing more control for operations involving skin shaping or cauterization.

The Air Force Office of Scientific Research is supporting CRI's research into applications, including materials for infrared waveguides and sensors needed to identify chemical components. CRI is also continuing basic research on fragile oxide liquids, which they believe still offer much potential for generating new materials and ultimately optical devices.

4.2.2 Optical Fibers in Sponges

Scientists at Lucent Technologies' Bell Labs have found that a deep-sea sponge contains optical fiber that is remarkably similar to the optical fiber found in today's state-of-the-art telecommunications networks. The deep-sea sponge's glass fiber, designed through the course of evolution, may possess certain technological advantages over industrial optical fiber. "We believe this novel biological optical fiber may shed light upon new bio-inspired processes that may lead to better fiber optic materials and networks," says Joanna Aizenberg, the Bell Labs materials scientist who led the research team. "Mother Nature's ability to perfect materials is amazing, and the more we study biological organisms, the more we realize how much we can learn from them."

The sponge in the study, *Euplectella*, lives in the depths of the ocean in the tropics and grows to about half a foot in length. Commonly known as the

Venus Flower Basket, it has an intricate cylindrical mesh-like skeleton of glassy silica, often inhabited by a pair of mating shrimp. At the base of the sponge's skeleton is a tuft of fibers that extends outward like an inverted crown. Typically, these fibers are between two and seven inches long and about the thickness of a human hair.

The Bell Labs team found that each of the sponge's fibers comprises distinct layers with different optical properties. Concentric silica cylinders with high organic content surround an inner core of high-purity silica glass, a structure similar to industrial optical fiber, in which layers of glass cladding surround a glass core of slightly different composition. The researchers found during experiments that the biological fibers of the sponge conducted light beautifully when illuminated and found them to use the same optical principles that modern engineers have used to design industrial optical fiber. "These biological fibers bear a striking resemblance to commercial telecommunications fibers, as they use the same material and have similar dimensions," says Aizenberg.

Although these natural bio-optical fibers do not have the superbly high transparency needed for modern telecommunication networks, the Bell Labs researchers found that these fibers do have a big advantage in that they are extremely resilient to cracks and breakage. Commercial optical fiber is extremely reliable; however, outages can occur mainly due to crack growth within the fiber. Infrequent as an outage is, when it occurs, replacing the fiber is often a costly, labor-intensive proposition, and scientists have sought to make fiber that is less susceptible to this problem.

The sponge's solution is to use an organic sheath to cover the biological fiber, Aizenberg and her colleagues discovered. "These bio-optical fibers are extremely tough," she says. "You could tie them in tight knots, and, unlike commercial fiber, they would still not crack. Maybe we can learn how to improve on existing commercial fiber from studying these fibers of the Venus Flower Basket."

Another advantage of these biological fibers is that they are formed by chemical deposition at the temperature of seawater. Commercial optical fiber is produced with the help of a high-temperature furnace and expensive equipment. Aizenberg says, "If we can learn from nature, there may be an alternative way to manufacture fiber in the future."

Should scientists succeed in emulating these natural processes, they may also help reduce the cost of producing optical fiber. "This is a good example where Mother Nature can help teach us about engineering materials," says Cherry Murray, senior vice president of physical sciences research at Bell Labs. "In this case, a relatively simple organism has a solution to a very complex problem in integrated optics and materials design. By studying the Venus Flower Basket, we are learning about low-cost ways of forming complex optical materials at low temperatures. While many years away from being applied to commercial use, this understanding could be very important in reducing the cost and improving the reliability of future optical and telecommunications equipment."

4.2.3 Mineral Wire

Researchers have developed a process to create wires only 50 nm (billionths of a meter) thick. Made from silica, the same mineral found in quartz, the wires carry light in an unusual way. Because the wires are thinner than the wavelengths of light they transport, the material serves as a guide around which light waves flow. In addition, because the researchers can fabricate the wires with a uniform diameter and smooth surfaces down to the atomic level, the light waves remain coherent as they travel.

The smaller fibers will allow devices to transmit more information while using less space. The new material may have applications in ever-shrinking medical products and tiny photonics equipment such as nanoscale laser systems, tools for communications, and sensors. Size is of critical importance to sensing—with more, smaller-diameter fibers packed into the same area, sensors could detect many toxins, for example, at once and with greater precision and accuracy.

Researchers at Harvard University led by Eric Mazur and Limin Tong (also of Zhejiang University in China), along with colleagues from Tohoku University in Japan, report their findings in the December 18, 2003, issue of *Nature*.

The NSF, a pioneer among federal agencies in fostering the development of nanoscale science, engineering, and technology, supports Mazur's work." Dr. Mazur's group at Harvard has made significant contributions to the fields of optics and short-pulse laser micromachining," says Julie Chen, program director of NSF's nanomanufacturing program. "This new method of manufacturing subwavelength-diameter silica wires, in concert with the research group's ongoing efforts in micromachining, may lead to a further reduction of the size of optical and photonic devices."

4.2.4 Hybrid Pastic

Leveraging their growing laser expertise, University of Toronto researchers have developed a hybrid plastic that can produce light at wavelengths used for fiber-optic communication, paving the way for an optical computer chip.

The material, developed by a joint team of engineers and chemists, is a plastic embedded with quantum dots—crystals just five billionths of a meter in size—that convert electrons into photons. The findings hold promise for directly linking high-speed computers with networks that transmit information using light.

"While others have worked in quantum dots before, we have shown how quantum dots can be tuned and incorporated into the right materials to address the whole set of communication wavelengths," says Winslow Sargeant,

NSF Program officer for small business. "Our study is the first to demonstrate experimentally that we can convert electrical current into light using a particularly promising class of nanocrystals." The research is based on nanotechnology: engineering based on the length of a nanometer—one billionth of a meter. "We are building custom materials from the ground up," says Winslow Sargeant, NSF Program officer for small business.

Working with colleagues in the university's chemistry department, the team created lead sulfide nanocrystals, using a cost-effective technique that allowed them to work at room pressure and at the relatively cool temperatures of less than 150 degrees Celsius. Traditionally, creating the crystals used in generating light for fiber-optic communications means working in a vacuum at temperatures approaching 600 to 800 degrees Celsius.

"Despite the precise way in which quantum dot nanocrystals are created, the surfaces of the crystals are unstable," says Gregory Scholes, a chemistry department professor. To stabilize the nanocrystals, the team encircled them with a special layer of molecules. The crystals were then combined with a semiconducting polymer material to create a thin, smooth film of the hybrid polymer.

Sargent explains that, when electrons cross the conductive polymer, they encounter what are essentially "canyons," with a quantum dot located at the bottom. Electrons must fall over the edge of the "canyon" and reach the bottom before producing light. The team tailored the stabilizing molecules so that they would hold special electrical properties, ensuring a flow of electrons into the light-producing "canyons."

The colors of light the researchers generated, ranging from 1.3 to 1.6 μm in wavelength, spanned the full range of colors for communicating information with the use of light. "Our work represents a step toward the integration of many fiber-optic communications devices on one chip," says Sargent. "We've shown that our hybrid plastic can convert electric current into light with promising efficiency and with a defined path toward further improvement. With this light source, combined with fast electronic transistors, light modulators, light guides, and detectors, the optical chip is in view."

4.2.5 Buckyballs

University of Toronto researchers are also looking into how to gain better control over light. Right now, managing light signals (photons) with electronic hardware is difficult and expensive, which makes it difficult to harness fast and free-flowing photons. Yet help may soon be on the way, however. That's because University of Toronto researchers have developed a new material that could make photon control less expensive and far easier. Using molecules resembling 60-sided soccer balls, the researchers—based at the University of Toronto and Carleton University—believe that the material will eventually give optical network users a powerful new way to process information using light.

Along with Sargent and Carleton University chemistry professor Wayne Wang, the team developed a material that combines microscopic spherical particles—known as "buckyballs"—with polyurethane, a polymer often used as a coating on cars and furniture.

Buckyballs, named after geodesic dome inventor Buckminster Fuller, are clusters of 60 carbon atoms, resembling soccer balls, that are only a few nanometers in diameter. Given the chemical notation C60, buckyballs were identified in 1985 by three scientists who later received a Nobel Prize for their discovery. Buckyballs have been the building block for many experimental materials and are widely used in nanotechnology research.

When a mixture of polyurethane and buckyballs is used as a thin film on a flat surface, light particles traveling though the material pick up each others' patterns. The new material has the potential to make the delivery and processing of information in fiber-optic communications more efficient. "In our high-optical-quality films, light interacts 10 to 100 times more strongly with itself, for all wavelengths used in optical fiber communications, than in previously reported C60-based materials," says Sargent. "We've also shown for the first time that we can meet commercial engineering requirements: the films perform well at 1,550 nm, the wavelength used to communicate information over long distances."

Creating the material required research that was not unlike assembling a complex jigsaw puzzle. "The key to making this powerful signal-processing material was to master the chemistry of linking together the buckyballs and the polymer," says Wang.

Although it will be several years before the new material can enter commercial use, its development proves an important point, says Sargent, "This work proves that 'designer molecules' synthesized using nanotechnology can have powerful implications for future generations of computing and communications networks."

4.2.6 Old Glass/New Promise

An Ohio State University engineer and his colleagues have discovered something new about a 50-year-old type of fiberglass: it may be more than one and a half times stronger than previously thought. That conclusion, and the techniques engineers used to reach it, could help expand applications for glass fibers.

The half-century-old glass, called E-glass, is the most popular type of fiberglass and is often used to reinforce plastic and other materials. Prabhat K. Gupta, a professor of materials science and engineering at Ohio State University and his coresearchers have developed an improved method for measuring the strength of E-glass and other glass fibers, including those used in fiber-optic communications. The method could lead to the development of stronger and cheaper fiber runs.

The measuring method would be relatively easy to implement in industry, since it only involves holding a glass fiber at low temperatures and bending it until it breaks. The key, Gupta says, is ensuring that a sample is completely free of flaws before the test. Gupta isn't surprised that no one has definitively measured the strength of fiberglass before now. "Industries develop materials quickly for specific applications," he says. "Later, there is time for basic research to further improve a material."

To improve a particular formulation of glass and devise new applications for it, researchers need to know how strong it is under ideal conditions. Therefore, Gupta and his colleagues—Charles Kurkjian, formerly of AT&T Bell Labs and now a visiting professor of ceramic and materials engineering at Rutgers University; Richard Brow, professor and chairman of ceramic engineering at University of Missouri-Rolla; and Nathan Lower, a masters student at University of Missouri-Rolla—had to determine the ideal conditions for the material.

In their latest work, the engineers outlined a set of procedures that researchers in industry and academia can follow to assure that they are measuring the ideal strength of a glass fiber. For instance, if small-diameter versions of the fiber seem stronger than larger-diameter versions, then the glass most likely contains flaws. That's because the ideal strength depends on inherent qualities of the glass, not the diameter of the fiber, Gupta says.

To measure the ideal strength of E-glass, Gupta and his coresearchers experimented on fibers that were 100 µm thick—about the same thickness as a human hair—held at minus 320°F. They bent single fibers into a "U" shape and pressed them between two metal plates until the fibers snapped at the fold. The fibers withstood a pressure of almost 1.5 million pounds per square inch—roughly 1.7 times higher than previously recorded measurements of 870,000 pounds per square inch. The results suggest that the engineers were able to measure the material's true strength.

Given the telecommunications industry's current slump, however, Gupta doubts that optical fiber makers will be looking to dramatically improve the strength of their product. "Even very high quality optical fiber is dirt cheap today," he says. "A more likely application is in the auto industry, where reinforced plastics could replace metal parts and make cars lighter and more fuel efficient."

Gupta and his colleagues next hope to study the atomic level structure of glass and learn more about what contributes to strength at that level.

4.3 NANOPHOTONICS

A Cornell University researcher is developing microscopic nanophotonic chips—which replace streams of electrons with beams of light—and ways of connecting the devices to optical fiber. Michal Lipson, an assistant professor at Cornell's School of Electrical and Computer Engineering, believes that one of the first applications of nanophotonic circuits might be as routers and repeaters for fiber-optic communication systems. Such technology, she notes, could speed the day when residential use of fiber-optic lines becomes practical.

Previous nanoscale photonic devices used square waveguides—a substitute for wiring—that confine light by total internal reflection. But this approach works only in materials with a high index of refraction, such as silicon, because these materials tend to reduce light intensity and distort pulses. Lipson has discovered a way to guide and bend light in low-index materials, including air or a vacuum. "In addition to reducing losses, this opens the door to using a wide variety of low-index materials, including polymers, which have interesting optical properties," Lipson says.

Using equipment at the Cornell Nanoscale Facility, Lipson's group has manufactured waveguides consisting of two parallel strips of a material with a high refractive index. The strips were placed about 50 to 200 nm apart, with a slot containing a material of much lower refractive index. In some devices, the walls are made of silicon with an air gap, whereas others have silicon dioxide walls with a silicon gap. In both cases, the index of refraction of the medium in the gap is much lower than that of the wall, up to a ratio of about four to one.

When a wavefront crosses two materials of very different refractive indices, and the low-index space is very narrow in proportion to the wavelength, nearly all of the light is confined in the "slot waveguide." Theory predicts that straight slots will have virtually no loss of light, and smooth curves will have only a small loss. This characteristic has been verified by experiments, Lipson reports.

Slot waveguides can be used to make ring resonators, which are already familiar to nanophotonics researchers. When a circular waveguide is placed very close to a straight one, some of the light can jump from the straight to the circular waveguide, depending on its wavelength. "In this way, we can choose the wavelength we want to transmit," Lipson says. In fiber-optic communications, signals often are multiplexed, with several different wavelengths traveling together in the same fiber and with each wavelength carrying a different signal. Ring resonators can be used as filters to separate these signals, Lipson notes.

Like the transistor switches in conventional electronic chips, light-beam switches would be the basic component of photonic computers. Lipson's group has made switches in which light is passed in a straight line through a cavity with reflectors at each end, causing the light to bounce back and forth many times before passing through. The refractive index of the cavity is varied by applying an electric field; because of the repeated reflections, the light remains in the waveguide long enough to be affected by this small change. Lipson is working on devices in which the same effect is induced directly by another beam of light.

Connecting photonic chips to optical fibers can be a challenge because the fiber is usually much larger than the waveguide, like trying to connect a garden hose to a hypodermic needle. Most researchers have used waveguides that taper from large to small, but the tapers typically have to be very long and thus introduce losses. Lipson's group, however, has made waveguides that narrow almost to a point. When light passes through the point, the waveform is deformed as if it were passing through a lens, spreading out to match the larger fiber. Conversely, the "lens" collects light from the fiber and focuses it into the waveguide. Lipson calls this coupling device "optical solder." According to experiments at Cornell, the device could couple 200-nm waveguides to 5-μm fibers with 95 percent efficiency. It can also be used to couple waveguides of different dimensions.

4.4 WAVE POLARIZATION

A new and novel way of communicating over fiber optics is being developed by physicists supported by the Office of Naval Research. Rather than using the amplitude and frequency of electromagnetic waves, they are using the polarization of the wave to carry the signal. Such a method offers a novel and elegant method of securing communications over fiber-optic lines.

Electromagnetic waves, like light and radio waves, have amplitude (wave height), frequency (how often the wave crests each second), and polarization (the plane in which the wave moves). Changes in amplitude and frequency have long been used to carry information. For example, AM radio uses changes in the amplitude of radio waves, whereas FM radio uses changes in frequency. Yet wave polarization has not been so thoroughly explored.

Office of Naval Research-supported physicists Gregory Van Wiggeren from the Georgia Institute of Technology, and Rajarshi Roy, from the University of Maryland, have demonstrated an ingenious method to communicate through fiber optics by using dynamically fluctuating states of light polarization. Unlike previous methods, the state of the light's polarization is not directly used to encode data. Instead the message (encoded as binary data of the sort used by digital systems) modulates a special kind of laser light. Van Wiggeren and Roy used an erbium-doped fiber ring laser. The erbium amplifies the optical signal, and the ring laser transmits the message. In a ring laser, the coherent laser light moves in a ring-shaped path, but the light can also be split from the ring to be transmitted through a fiber optic cable.

The nonlinearities of the optic fiber produce dynamical chaotic variations in the polarization, and the signal is input as a modulation of this naturally occurring chaos. The signal can be kept small relative to the background light amplitude. The light beam is then split, with part of it going through a communications channel to a receiver. The receiver breaks the transmitted signal into two parts. One of these is delayed by about 239 nanoseconds, the time it takes the signal to circulate once around the ring laser. The light received directly is compared, by measuring polarizations, with the time-delayed light. The chaotic variations are then subtracted, which leaves only the signal behind. Variations in stress and temperature on the communications would be equally subtracted out.

"This is quite a clever method, which hides the signal in noise," says Office of Naval science officer Mike Shlesinger, who oversees the research. "It provides a definite advantage over direct encoding of polarization, leaving an eavesdropper only chaotic static and no means to extract the signal."

4.5 OPTICAL COMMUNICATIONS VIA CDMA

A new generation of light-based communications devices is the goal of engineers at the University of California, Davis, and the Lawrence Livermore National Laboratory. The researchers will build chip-sized devices that use code division multiple access (CDMA), a method already in use in some mobile phones, to transmit and receive optical signals.

Optical fibers and lasers send messages as streams of light pulses. In current technology, different messages are separated with a method called wavelength division multiplexing (WDM), where each message uses a different wavelength of light. In contrast, optical CDMA encodes each pulse or bit of information across a spread of wavelengths. The receiver uses a key to decode the signal and recreate the original pulse. "You don't need a wavelength for each user," says Ben Yoo, an associate professor of electrical engineering at UC Davis.

Optical CDMA devices would provide fast, secure links to telecommunications networks that already use an optical fiber backbone, Yoo notes. Currently, users access the network backbone over slower electronic or wireless connections. CDMA also makes it difficult for eavesdroppers to tune in as the frequencies used change rapidly. Even if a snoop can tap into a conversation, they can't understand it without knowing the key. "Security-wise, there are strong advantages to optical CDMA because you can change the code at any time," says UC Davis electrical engineer Zhi Ding.

4.6 LIGHT EMITTERS

Optical networks are all about light. So it's only natural that researchers are working hard to create light sources that are bright, smaller, and less power hungry.

4.6.1 Smallest Light Emitter

IBM researchers have created the world's smallest solid-state light emitter. The device, the first electrically controlled, single-molecule light emitter—demonstrates the rapidly improving understanding of molecular devices. The results also suggest that the unique attributes of carbon nanotubes may be applicable to optoelectronics, which is the basis for the high-speed communications industry.

IBM's previous work on the electrical properties of carbon nanotubes has helped establish carbon nanotubes as a top candidate to replace silicon when current chip features can't be made any smaller. Carbon nanotubes are tube-shaped molecules that are 50,000 times thinner than an average human hair. IBM scientists expect today's achievement to spark additional research and interest in the use of carbon nanotubes in nanoscale electronic and photonic (light-based) devices.

"By further understanding the electrical properties of carbon nanotubes through their light emission, IBM is accelerating the development path for their electronic applications, as well as possible optical applications," says Dr. Phaedon Avouris, manager of nanoscale science, IBM Research. "Nanotube light emitters have the potential to be built in arrays or integrated with carbon nanotube or silicon electronic components, opening new possibilities in electronics and optoelectronics."

IBM's research team has detected light with a wavelength of 1.5 μm, which is particularly valuable because it is the wavelength widely used in optical communications. Nanotubes with different diameters could generate light with different wavelengths used in other applications.

IBM's light emitter is a single nanotube, 1.4 nm in diameter, configured into a three-terminal transistor. As in a conventional semiconductor transistor, applying a low voltage to the transistor's gate switches current passing between opposite ends of the nanotube (the device's source and drain).

Building on their previous research, IBM scientists engineered the device to be "ambipolar," so they could simultaneously inject negative charges (electrons) from a source electrode and positive charges (holes) from a drain electrode into a single carbon nanotube. When the electrons and holes meet in the nanotube, they neutralize each other and generate light.

Because it is a transistor, IBM's light emitter can be switched on and off depending on the voltage applied to the gate of the device. Electrical control of the light emission of individual nanotubes allows detailed investigations of the optical physics of these unique one-dimensional materials. IBM researchers compared the characteristics of the emitted light with theoretical predictions to prove that the light was created by the electron-hole recombination mechanism.

Although optical emission from individual molecules has been measured before, that light emission was induced by laser irradiation of samples of molecules. In the case of carbon nanotubes, light emission is from a collection of a large number of nanotubes suspended in a liquid irradiated with a laser.

4.6.2 Light-Emitting Transistor

Researchers at the University of Illinois-Urbana/Champaign have created a light-emitting transistor that could make the transistor the fundamental element in optoelectronics as well as in electronics (Fig. 4-2).

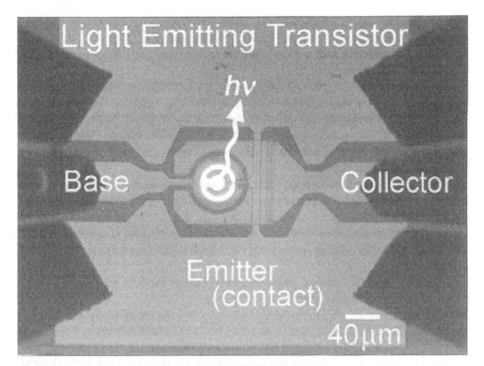

Figure 4-2 *Light-emitting transistor.*

"We have demonstrated light emission from the base layer of a hetero-junction bipolar transistor and showed that the light intensity can be controlled by varying the base current," says Nick Holonyak, a John Bardeen professor of electrical and computer engineering and physics at the University of Illinois. Holonyak invented the first practical light-emitting diode and the first semiconductor laser to operate in the visible spectrum. "This work is still in the early stage, so it is not yet possible to say what all the applications will be," says Holonyak. "But a light-emitting transistor opens up a rich domain of integrated circuitry and high-speed signal processing that involves both electrical signals and optical signals."

A transistor usually has two ports: one for input and one for output. "Our new device has three ports: an input, an electrical output, and an optical output," says Milton Feng, professor of electrical and computer engineering at Illinois. "This means that we can interconnect optical and electrical signals for display or communication purposes." Feng is credited with creating the world's fastest bipolar transistor, a device that operates at a frequency of 509 GHz.

Graduate student Walid Hafez fabricated the light-emitting transistor at the University of Illinois' micro and nanotechnology lab. Unlike traditional transistors, which are built from silicon and germanium, the light-emitting transistors are made from indium gallium phosphide and gallium arsenide. "In a bipolar device, there are two kinds of injected carriers: negatively charged electrons and positively charged holes," Holonyak says. "Some of these carriers will recombine rapidly, supported by a base current that is essential for the normal transistor function."

The recombination process in indium gallium phosphide and gallium arsenide materials also creates infrared photons, the "light" in the researchers' light-emitting transistors. "In the past, this base current has been regarded as a waste current that generates unwanted heat," says Holonyak. "We've shown that for a certain type of transistor, the base current creates light that can be modulated at transistor speed."

Although the recombination process is the same as that which occurs in light-emitting diodes, the photons in light-emitting transistors are generated under much higher speed conditions. So far, the researchers have demonstrated the modulation of light emission in phase with a base current in transistors operating at a frequency of 1 MHz. Much higher speeds are considered certain.

"At such speeds, optical interconnects could replace electrical wiring between electronic components on a circuit board," says Feng. This work could be the beginning of an era in which photons are directed around a chip in much the same fashion as electrons have been maneuvered on conventional chips.

"In retrospect, we could say the groundwork for this was laid more than 56 years ago with John Bardeen and Walter Brattain and their first germanium transistor," says Holonyak, who was Bardeen's first graduate student. "But the direct recombination involving a photon is weak in germanium materials, and John and Walter just wouldn't have seen the light—even if they had looked. If John were alive and we showed him this device, he would have to have a big grin."

4.6.3   VCSEL

A VCSEL (pronounced "vixel") is a tiny laser that emits a beam vertically from a chip's surface. PARC has developed the world's most densely packed VCSEL array, with independently addressable laser elements placed on a linear pitch as small as 3 μm.

Silicon CMOS-based driver-integrated circuits can be combined with the laser array to form an ultradense optoelectronic multi-chip module using PARC's StressedMetal MEMS high-density interconnect technology. Such an imager-on-a-chip allows thousands of independently addressable lasers to be packaged in a compact module. The imagers could eliminate optical systems with mechanical moving parts, replacing them with a more compact, all solid-state light source.

VCSELs could be used for building optical interconnects that transfer data between computers via lasers. One application would replace copper cables with bundles of optical fiber ribbons to connect electronic boards in computer back-planes, enabling much faster data transfer rates.

4.6.4 Improved VCSEL

With optical communications becoming increasingly important to telecom manufacturers, service providers, and users, scientists are pushing hard to create better performing lasers that promise to increase distance, boost efficiency, and reduce the cost of optical connections. At the University of Illinois at Urbana-Champaign, researchers have found a way to significantly improve the performance of VCSEL by drilling holes in their surfaces. Faster and cheaper long-haul optical communication systems, as well as photonic integrated circuits, could be the result.

Low-cost VCSELs are currently used in data communication applications where beam quality is of little importance, such as in short fiber runs. To operate at higher speeds and over longer distances, however, the devices must function in a single transverse mode with a carefully controlled beam. "These characteristics are normally found only in very expensive lasers, not in mass-produced VCSELs," says Kent D. Choquette, a professor of electrical and computer engineering and a researcher at the university's Micro and Nanotechnology Laboratory. "By embedding a two-dimensional photonic crystal into the top face of a VCSEL, however, we can accurately design and control the device's mode characteristics."

The two-dimensional photonic crystal, created by drilling holes in the semiconductor surface, introduces a periodic change in the index of refraction, Choquette says. The holes represent regions of low refractive index, surrounded by semiconductor material where the index is higher. A particular combination of refractive indices will produce a single-mode waveguide that permits only one transverse wave of the laser beam to propagate. "Our photonic crystal consists of a triangular array of circular holes that have been etched into the top of a VCSEL," Choquette says. "Because the index variation has to be on the length scale of light, the periodicity of the holes must be on the order of several hundred nanometers."

To create such a precise array of holes, the researchers first lithographically define the desired pattern into a silicon dioxide mask layer on the semiconductor surface using focused-ion beam etching. Researchers then bore holes into the semiconductor material using inductively coupled plasma etching. "By selectively varying parameters such as depth, diameter, and spacing of the holes, we can control the modal characteristics of the laser," Choquette says. "This means we can accurately design and fabricate single-mode VCSELs for high-performance optical communication systems."

The next step, he says, is to push VCSEL performance toward higher power by considering designs that are much larger in diameter. "Looking beyond that, we also have fundamental problems with high-speed data communication on our circuit boards and in our chips," Choquette says. "This is a technology that could serve as the foundation for a new way of looking at optical interconnects and photonic integrated circuits."

4.6.5 Tiny Laser

Bell Labs researchers have built a tiny semiconductor laser that has the potential for advanced optical communication applications, including multiple laser-on-a-chip applications. The device exploits a photonic crystal, a highly engineered material with superior optical properties, and was made in collaboration with scientists from the New Jersey Nanotechnology Consortium, California Institute of Technology, and Harvard University.

"This new laser was made possible by taking advances from many different areas in physics and incorporating them into one device," says Cherry Murray, senior vice president of physical sciences research at Bell Labs. "This work will open up new directions not only for optoelectronics and sensing but could also provide a new tool to investigate very basic physical phenomena."

The laser belongs to a class of high-performance semiconductor lasers, known as quantum cascade (QC) lasers, which were invented at Bell Labs in 1994. QC lasers are made by stacking many ultrathin atomic layers of standard semiconductor materials (such as those used in photonics) on top of each another, much like a club sandwich. By varying the thickness of the layers, it's possible to select the particular wavelength at which a QC laser will emit light, allowing scientists to custom design a laser.

When an electric current flows through a QC laser, electrons cascade down an energy "staircase," and every time an electron hits a step, a photon of infrared light is emitted. The emitted photons are reflected back and forth inside the semiconductor resonator that contains the electronic cascade, stimulating the emission of other photons. This amplification process enables high-output power from a small device.

In the decade since their invention, QC lasers have proven to be very convenient light sources. Besides being powerful, they are compact, rugged, and often portable. Yet the devices can't emit light from its edges or through its surface.

The Bell Labs team overcame this challenge by using the precise light-controlling qualities of a photonic crystal to create a QC laser that emits photons perpendicular to the semiconductor layers, resulting in a laser that emits light through its surface. Photonic crystals are materials with repeating patterns spaced very close to one another, with separations between the patterns comparable to the wavelengths of light. When light falls on such a patterned material, the photons of light interact with it; with proper design of the patterns, it's possible to control and manipulate the propagation of light within the material.

Using a state-of-the-art electron beam lithography facility at the New Jersey Nanotechnology Consortium, located at Bell Labs headquarters in Murray

Hill, New Jersey, the researchers were able to superimpose a hexagonal photonic crystal pattern on the semiconductor layers that made the QC laser. The final laser was only 50 μm across or about half the diameter of a human hair.

"The most exciting part of this work is that we combined photonic and electronic engineering to create a new surface-emitting QC laser," says Al Cho, adjunct vice president of semiconductor research at Bell Labs and one of the inventors of the QC laser. "The photonic crystal approach has real potential for new applications. The production of surface-emitting compact lasers only 50 micrometers across enables large arrays of devices to be produced on a single chip, each with its own designed emission properties."

Such lasers-on-chips may lead to new possibilities for optical communications, as well as other optoelectronic and sensing technologies. QC lasers have already been used to make extremely sensitive sensors, including sensors that have been used by NASA for atmospheric monitoring.

"The next step is to see if we can use this sort of technique to get sensing done within the laser," says Federico Capasso, a Harvard University consultant who is also a Bell Labs consultant and one of the inventors of the QC laser. "If we can fill the holes of the photonic crystal in this laser with nanoliters of fluid or other special material, we may get some interesting physics as well as a whole new world of applications."

4.6.6 Looking Into Lasers

University of Toronto researchers are looking to create more powerful and efficient lasers for fiber-optic communication systems by looking inside lasers while they are operating. "We've seen the inner workings of a laser in action," says investigator Ted Sargent, a professor in the university's department of electrical and computer engineering. "We've produced a topographical map of the landscape that electrons see as they flow into these lasers to produce light." He says the findings could influence laser design, change the diagnosis of faulty lasers, and potentially reduce manufacturing costs.

Lasers are created by growing a complex and carefully designed series of nanometer-sized layers of crystals on a disk of semiconductor material known as a wafer. Ridges are etched into the crystal surface to guide laser light, thin metal layers are added on top and bottom, and the wafer is then cut into tiny cubes or chips. During the laser's operation, an electrical current flows into the chip, providing the energy to generate intense light at a specific wavelength used in fiber-optic communications.

The University of Toronto researchers focused their efforts on the "beating heart" portion of the laser (called the active region), where electronic energy is converted into light. Using a technique called scanning voltage microscopy, they examined the surface of an operating laser, picking up differences in voltage. These differences translate to a topographical image of the laser's energy surface, allowing researchers to visualize the forces an electron experiences along its path into the active region, Sargent says.

The team used its newly acquired information about the inside operations of the laser to determine the fraction of electric current that contributes to producing light. The balance of electrons are undesirably diverted from the active region: such current leakage wastes electrons and heats the device up, degrading performance. "We used direct imaging to resolve a contentious issue in the field: the effectiveness of electronic funneling into the active region of a ridge-waveguide laser," says Dayan Ban, a University of Toronto doctoral candidate who made the measurements. "Previously, uncorroborated models had fueled speculation by yielding divergent results. Now we know where the electrons go." Ban is now a researcher at the Institute for Microstructural Sciences of the National Research Council of Canada.

"Direct imaging of the functions that drive the action of a living laser could transform how we think about laser 'diagnosis and therapy,'" says Sargent, referring to the measurement and optimization of laser structures and their determination of the devices' inner workings. Designers now use a variety of computer simulations to model how lasers work, but research at the University of Toronto may show which simulations are the most accurate design tools. "With accurate models," says Sargent, "the designs we can create are more likely to result in devices that meet design requirements."

The research could also help in diagnosing faulty lasers. "If a particular laser fails, the kind of measurements that we are taking could provide some idea of why it failed and the design could then be modified," says St. John Dixon-Warren, a project coinvestigator and a physical chemist at Bookham Technology, an optical components manufacturer based in the United Kingdom.

Sargent says the findings could have larger implications for the creation of optical circuits for fiber-optic communications. "If we could fully develop these models and fully understand how lasers work, then we could start to build optical circuits with confidence and high probability of success," he says. "Optical chips akin to electronic integrated circuits in computers must be founded on a deep and broad understanding of the processes at work inside current and future generations of lasers."

4.6.7 Manipulating Light

Manipulating light on the nanoscale level can be a Herculean task, since the nanoscale level is so incredibly tiny—less than one-tenth the wavelength of light. Researchers at Argonne National Laboratory are making strides toward understanding and manipulating light at the nanoscale level by using the unusual optical properties of metal nanoparticles, opening the door to microscopic-sized devices such as optical circuits and switches.

Metal nanoparticles, such as extremely tiny spheres of silver or gold, can concentrate large amounts of light energy at their surfaces. The light energy confined near the surface is known as the near field, whereas ordinary light is known as far field. Many scientists believe that, by understanding how to manipulate near-field light, new optical devices could be built at dimensions far smaller than is currently possible. In an effort to characterize near-field behavior, a joint experimental and theoretical study published in the December 25, 2003, issue of the *Journal of Physical Chemistry B* used powerful high-resolution imaging and modeling techniques to detail how light is localized and scattered by metal nanoparticles.

Technologies, such as high-speed computers and internet routers, rely heavily on electrons flowing through wires to function. However, with the ever-increasing demand for higher data rates and smaller sizes, the complexity of electrical circuits becomes untenable. This challenge can be overcome by replacing electrons with photons (units of light), since the wave-like character of photons would reduce obstacles such as heat and friction within a given system. "In a nutshell, photons move faster than electrons," says experimental team leader Gary Wiederrecht. "They are a highly efficient power source just waiting to be harnessed."

"Using experimental and theoretical approaches, we were able to observe the interaction of light with the surfaces of the metal nanoparticles. We hope that this study will lead to the creation of optical technologies that can manipulate light with precision at nanoscale dimensions," explains Stephen Gray, lead theoretician of the *Journal* study.

To obtain a more comprehensive understanding of the near field, the Argonne researchers used an advanced imaging technique known as near-field scanning optical microscopy. The nanoparticles, with diameters as small as 25 nm, were placed on a prism and illuminated with laser light, forming a near field that was detectable with near-field scanning optical microscopy by a nanoscale probe positioned close to the sample's surface. Optical scattering experiments were performed on isolated metal nanoparticles and arrays of metal nanoparticles. Electron beam lithography was used to uniformly place nanoparticles within 100 nm from one another. Using a special experimental setup, the team was able to explicitly map the near-field light intensity onto the three-dimensional topography of the metal nanoparticle arrays.

Experimental results yielded a number of valuable findings regarding the character of the near field. The researchers found that an isolated nanoparticle would scatter light at a 20-degree angle from the prism surface. Furthermore, the researchers found that arrays of nanoparticles scatter light at much smaller angles, an encouraging result for the use of near-field photons in two-dimensional devices such as optical chips. All findings were validated by computational and theoretical methods; together, there findings provide specific information as to how near fields can be used to guide light.

4.7 OPTICAL ANTENNA

A new optical antenna, developed by researchers at the University of Warwick in the United Kingdom, promises to bring significant benefits to wireless networks, household electronics, and long-distance data transfers. The device applies techniques used to manipulate radio frequencies to select the incoming "signal frequencies" carried on infrared beams to produce the optical equivalent of the radio.

The antenna, developed by Professor Roger Green and Roberto Ramirez-Iniguez, in the University of Warwick's engineering department, has been licensed to Optical Antenna Solutions of Nottingham, United Kingdom, which will be responsible for further development. Derick Wilson, Optical Antenna Solutions' managing director, believes the technology provides enormous benefits to several industries and has significant commercial opportunities. One of the first applications being considered is to use the antenna to provide secure practical "point and pay methods" for credit cards.

An optical antenna serves the same purpose as an electronic one. It must collect energy from an area and channel it through to a receiving element or, conversely, transmit energy that has originated from a small source over an area. The new device uses a combination of precise curvatures on the lens part of the instrument with a multi-layered filter. The optical antenna is so precise that it can detect a signal on one particular wavelength of light; it is also 100 times more efficient at gathering a signal than any previous optical sensor of this kind. This has immediate benefits for indoor wireless networks and household devices. It allows signal transmitters and receivers to operate at a significant angle to each other and to discriminate much more finely between signals. For external data transfer applications it can be used to greater distances—up to 3 miles.

Two major transmission technologies are presently used to achieve indoor wireless communication: radio and infrared. For many reasons, infrared is often preferred. Infrared links provide high bandwidth at low cost, infrared is immune to radio interference, the spectrum is freely available, and infrared components are inexpensive, small, and consume little power. The new optical antenna could help turn even more people toward infrared as an alternative to the high cost of maintaining wired networks.

4.8 KEEPING COPPER

With all of the ongoing research into optical technologies, it's easy to forget that old-fashioned copper wiring remains a perfectly suitable, and cheap, technology for many telecommunication applications, particularly those that involve short distances. Penn State University engineers have developed a copper wire transmission scheme for distributing a broadband signal over local area networks (LANs) with a lower average bit error rate than fiber optic cable that is 10 times more expensive.

"Using copper wire is much cheaper than fiber-optic cable and, often, the wire is already in place," says Mohsen Kavehrad, a Penn State professor of electrical engineering and director of the school's Center for Information and

Communications Technology Research. "Our approach can improve the capability of existing local area networks and shows that copper is a competitor for new installations in the niche LAN market."

The Penn State approach was developed in response to an IEEE challenge. The organization wanted a signaling scheme for a next generation broadband copper Ethernet network capable of carrying broadband signals of 10 gigabits per second. Currently, the IEEE standard specifies 1 gigabit over 100 meters of category 5 copper wire, which has four twisted pairs of wire in each cable.

"In the existing copper gigabit systems, each pair of wires carries 250 megabits per second. For a 10-gigabit system, each pair will have to carry 2.5 gigabits per second," says Kavehrad. "At these higher speeds, some energy penetrates into the other wires and produces crosstalk." The Penn State scheme eliminates crosstalk by using a new error correction method that they developed that jointly codes and decodes the signal and, in decoding, corrects the errors.

"Conventional wisdom says you should deal with the wire pairs one pair at a time, but we look at them jointly," Kavehrad says. "We use the fact that we know what signal is causing the crosstalk interference because it is the strongest signal on one of the wires." The Penn State approach also accounts for the reduction or loss of signal energy between one end of the cable and the other, which can become severe in 100-meter copper systems.

"We jointly code and decode the signals in an iterative fashion and, at the same time, we equalize the signals," Kavehrad says. "The new error correction approach acts like a vacuum cleaner where you first go over the rough spots and then go back again to pick up more particles."

A simulation showed that the scheme is possible and can achieve an average bit error rate of 10^{-12} bits per second. Fiber-optic cable typically achieves 10^{-9} bits per second. The work is continuing.

Chapter 5

The Internet Rules— IP Technologies

It's hard to believe that just 10 years ago only a select group of people (mostly scientists and military leaders) was aware of the Internet. Today, it's difficult to imagine a world without Web surfing, e-mail, online chats, or (sad to say) spam. Tomorrow's Internet may still be plagued by spam in one form or another (some miseries never seem to go away), but at least data will be faster and more multimedia oriented.

5.1 VoIP TELEPHONY

Perhaps the biggest impact the Internet will have on telecom will come from voice-over Internet protocol (VoIP) technology. Computer-based telephony is a key technology that promises to supplement and perhaps even eventually replace conventional telephones. VoIP technology has been around for several years but has so far failed to live up to its potential. Still, a study by Parks Associates, a technology research firm located in Dallas, Texas, finds that one-half of all Internet households are interested in VoIP services, which could deprive current long-distance companies of their most profitable subscribers.

Interest in these services is disproportionately high among subscribers with oppressive monthly phone bills, with cost savings being the primary driver. Furthermore, according to Parks' research, interest in VoIP service is consistent among both broadband (52 percent) and narrowband (48 percent) house-

Telecosmos: The Next Great Telecom Revolution, edited by John Edwards
ISBN 0-471-65533-3 Copyright © 2005 by John Wiley & Sons, Inc.

holds. "Consumers are looking at VoIP service as a replacement for their primary phone line, not as a secondary backup," says John Barrett, a research analyst at Parks. "VoIP could also serve as a lure for new broadband subscribers, given the strong interest among narrowband households. The market is still searching for a 'killer' broadband application. Wouldn't it be ironic if it turned out to be telephone service?"

Preconceived notions helped diminish VoIP's consumer potential. "When people think of consumer IP telephony, they often think of sound quality equal to two cans tied together with string or as a dot.com fad," says Daryl Schoolar, a senior analyst with In-Stat/MDR, a technology research firm based in Scottsdale, Arizona.

Still, although VoIP has been slow to gain traction with consumers, the technology is getting a second chance at life with enterprise adopters. In its enterprise incarnation, VoIP is experiencing solid growth. Probe Group, a telecom research company located in Cedar Knolls, New Jersey, forecasts that business-related IP voice services will experience geometric-level growth over the next several years. "While some of this growth will be derived from gradual increases in revenue per customer, the vast majority will come from rapid growth in market penetration," says Christine Hartman, a Probe analyst.

A combination of more widely dispersed organizations and shrinking travel/relocation budgets will help grow the VoIP market. Improved software and related technologies will also help boost the technology's acceptance. Additionally, vendors of hosted IP voice equipment have been refining and improving the reliability of their products, leading to an array of reliable solutions. Hosted solutions are particularly attractive in the present economic climate because businesses are able to access advanced features without the necessary investment in premise equipment.

Large carriers have addressed the market by providing managed services directed toward major enterprise customers. Yet smaller carriers are also developing hosted solutions, focusing on wholesale services and retail strategies that concentrate on specific geographic areas. The combined worldwide revenues for hosted VoIP services, including IP-PBX, videoconferencing, contact center, and unified communications, are expected to grow from $46 million in 2001 to $36.5 billion in 2008, according to research from ABI, a technology research firm based in Oyster Bay, New York.

"In 2003, we saw continued growth from emerging services providers who, after quietly launching their offers during 2001 and 2002, have been gradually building their customer base," says Julia Mermelstein, an ABI analyst. "While these providers gain customer traction in focused vertical or geographic markets, we will also see several incumbent telecom providers, who have been slow to engage in the hosted VoIP services market, convert trial activities into large-scale product launches."

Enhanced technologies and growing enterprise interest in VoIP are also renewing hope that the technology may finally achieve success in the consumer market. The key to any future consumer growth will be the enhanced quality provided by broadband Internet connections. "Overall, by 2007, the U.S. IP telephony market is forecast to grow to over 5 million active subscribers," says In-Stat/MDR's Schoolar. "While this shows a fivefold increase in subscribers over 2002, it still lags U.S. plain old telephone service (POTS) with over 100 million households."

As interest in VoIP grows, the private line business suffers. Once the cash cow of data transport, the future of private line services is in jeopardy as the world slowly migrates to IP. Whereas the bottom hasn't yet fallen out on private line expenditures (U.S. businesses spent roughly $23 billion on private line services in 2003, up about 4 percent), In-Stat/MDR expects that growth will stagnate in the near term, with this market facing an eventual and pronounced decline in the long term.

"The reality is that the public network is migrating to IP, meaning traditional circuit-switch private lines will need to migrate as well," says Kneko Burney, In-Stat/MDR's chief market strategist. Today, the most common migration path is primarily to switch to a high-end version of DSL, such as HDSL, and this is likely to escalate over time as DSL reach and capabilities broaden. "However, this migration will be gradual," says Burney, "meaning that the T1 businesses will experience a long, slow, and possibly painful exit as replacement escalates—similar to that experienced by long distance and now, local phone service."

Burney believes that traditional T1 providers may be able to manage the erosion through innovation—meaning stepping up plans to offer integrated T1 lines—and by focusing on specific segments of the market, like midsized businesses (those with 100 to 999 employees). "According to In-Stat/MDR's research, respondents from midsized businesses were somewhat less likely than their peers from both smaller and larger firms to indicate that they were planning or considering switching from T1 to integrated T1, cable, or S, H, or VDSL alternatives," says Colin Nelson, an In-Stat/MDR research analyst.

5.2 THE NEXT INTERNET

Today's Internet is simply amazing, particularly when combined with broadband access. Yet speeds are set to rise dramatically.

Organizations such as the academic-sponsored Internet2 and the U.S. government's Next Generation Internet are already working on developing a global network that can move information much faster and more efficiently than today's Internet. In 2003, Internet2 researchers sent data at a speed of 401 megabits per second (Mbps) across a distance of over 7,600 miles, effectively transmitting the contents of an entire CD in less than two minutes and providing a taste of what the future Internet may be like.

By 2025, we'll likely be using Internet version 3, 4, or 5 or perhaps an entirely new type of network technology that hasn't yet been devised. How fast will it run? Nobody really knows right now, but backbone speeds of well over 1 billion bps appear likely, providing ample support for all types of multimedia content. Access speeds, the rate at which homes and offices connect to the Internet, should also soar—probably to well over 100 Mbps. That's more than enough bandwidth to support text, audio, video, and any other type of content that users will want to send and receive.

The next-generation Internet will even revolutionize traditional paper-based publishing. Digital paper—thin plastic sheets that display high-resolution text and graphic images—offers the prime attributes of paper, including portability, physical flexibility, and high contrast, while also being reusable. With a wireless connection to the Internet, a single sheet of digital paper would give users access to an entire library of books and newspapers.

Ultimately, however, an ultrabroadband Internet will allow the creation of technologies that can't even be imagined today. Twenty years ago, nobody thought that the Internet would eventually become an everyday consumer technology. In the years ahead, the Internet itself may spin off revolutionary, life-changing "disruptive" technologies that are currently unimaginable. "It's very hard to predict what's going to be next," says Krisztina Holly, executive director of the Deshpande Center for Technological Innovation at the Massachusetts Institute of Technology. "Certainly the biggest changes will be disruptive technologies."

5.2.1 Riding the LambdaRail

An experimental new high-speed computing network—the National LambdaRail (NLR)—will allow researchers nationwide to collaborate in advanced research on topics ranging from cancer to the physical forces driving hurricanes.

The NLR consortium of universities and corporations, formed over the past several months, is developing a network that will eventually include 11,000 miles of high-speed connections linking major population areas. The LambdaRail name combines the Greek symbol for light waves with "rail," which echoes an earlier form of network that united the country.

NLR is perhaps the most ambitious research and education networking initiative since ARPANET and NSFnet, both of which eventually led to the commercialization of the Internet. Like those earlier projects, NLR is designed to stimulate and support innovative network research to go above and beyond the current Internet's incremental evolution.

The new infrastructure offers a wide range of facilities, capabilities, and services in support of both application level and networking level experiments. NLR will serve a diverse set of communities, including computational scientists, distributed systems researchers, and networking researchers. NLR's goal is to bring these communities closer together to solve complex architectural and end-to-end network scaling challenges.

Researchers have used the recently created Internet2 as their newest super-highway for high-speed networking. That system's very success has given rise to the NLR project. "Hundreds of colleges, universities, and other research institutions have come to depend on Internet2 for reliable high-speed transmission of research data, video conferencing, and coursework," says Tracy Futhey, chair of NLR's board of directors and vice president of information technology and chief information officer of Duke University. "While Internet2's Abilene network supports research, NLR will offer more options to researchers. Its optical fiber and light waves will be configured to allow essentially private research networks between two locations.

The traffic and protocols transmitted over NLR's point-to-point infrastructure provide a high degree of security and privacy. "In other words, the one NLR network, with its 'dark fiber' and other technical features gives us 40 essentially private networks, making it the ideal place for the sorts of early experimentation that network researchers need to develop new applications and systems for sharing information," says Futhey.

NLR is deploying a switched Ethernet network and a routed IP network over an optical DWDM network. Combined, these networks enable the allocation of independent, dedicated, deterministic, ultra-high-performance network services to applications, groups, networked scientific apparatus and instruments, and research projects. The optical waves enable building networking research testbeds at switching and routing layers with ability to redirect real user traffic over them for testing purposes. For optical layer research testbeds, additional dark fiber pairs are available on the national footprint.

NLR's optical and IP infrastructure, combined with robust technical support services, will allow multiple, concurrent large-scale networking research and application experiments to coexist. This capability will enable network researchers to deploy and control their own dedicated testbeds with full visibility and access to underlying switching and transmission fabric.

NLR's members and associates include Duke, the Corporation for Education Network Initiatives in California, the Pacific Northwest Gigapop, the Mid-Atlantic Terascale Partnership and the Virginia Tech Foundation, the Pittsburgh Supercomputing Center, Cisco Systems, Internet2, the Georgia Institute of Technology, Florida LambdaRail, and a consortium of the Big Ten universities and the University of Chicago.

Big science requires big computers that generate vast amounts of data that must be shared efficiently, so the Department of Energy's Office of Science has awarded Oak Ridge National Laboratory (ORNL) $4.5 million to design a network up to the task.

"Advanced computation and high-performance networks play a critical role in the science of the 21st century because they bring the most sophisticated scientific facilities and the power of high-performance computers literally to the researcher's desktop," says Raymond L. Orbach, director of the Department of Energy's science office. "Both supercomputing and high-performance networks are critical elements in the department's 20-year facilities plan that Secretary of Energy Spencer Abraham announced November 10th."

The prototype-dedicated high-speed network, called the Science UltraNet, will enable the development of networks that support high-performance computing and other large facilities at DOE and universities. The Science UltraNet will fulfill a critical need because collaborative large-scale projects typical today make it essential for scientists to transfer large amounts of data quickly. With today's networks, that is impossible because they do not have adequate capacity, are shared by many users who compete for limited bandwidth, and are based on software and protocols that were not designed for petascale data.

"For example, with today's networks, data generated by the terascale supernova initiative in two days would take two years to transfer to collaborators at Florida Atlantic University," says Nageswara Rao of Oak Ridge National Laboratory's Computer Science and Mathematics Division.

Obviously, Rao says, this is not acceptable; thus, he, Bill Wing, and Tom Dunigan of ORNL's Computer Science and Mathematics Division are heading the three-year project that could revolutionize the business of transferring large amounts of data. Equally important, the new UltraNet will allow for remote computational steering, distributed collaborative visualization, and remote instrument control. Remote computational steering allows scientists to control and guide computations being run on supercomputers from their offices.

"These requirements place different types of demands on the network and make this task far more challenging than if we were designing a system solely for the purpose of transferring data," Rao says. "Thus, the data transmittal requirement plus the control requirements will demand quantum leaps in the functionality of current network infrastructure as well as networking technologies."

A number of disciplines, including high-energy physics, climate modeling, nanotechnology, fusion energy, astrophysics, and genomics will benefit from the UltraNet.

ORNL's task is to take advantage of current optical networking technologies to build a prototype network infrastructure that enables development and testing of the scheduling and signaling technologies needed to process requests from users and to optimize the system. The UltraNet will operate at 10 to 40 Gbps, which is about 200,000 to 800,000 times faster than the fastest dial-up connection of 56,000 bps.

The network will support the research and development of ultra-high-speed network technologies, high-performance components optimized for very large-scale scientific undertakings. Researchers will develop, test, and optimize networking components and eventually make them part of Science UltraNet.

"We're not trying to develop a new Internet," Rao says. "We're developing a high-speed network that uses routers and switches somewhat akin to phone companies to provide dedicated connections to accelerate scientific discoveries. In this case, however, the people using the network will be scientists who generate or use data or guide calculations remotely."

The plan is to set up a testbed network from ORNL to Atlanta, Chicago, and Sunnyville, California. "Eventually, UltraNet could become a special-purpose network that connects DOE laboratories and collaborating universities and institutions around the country," Rao says. "And this will provide them with dedicated on-demand access to data. This has been the subject of DOE workshops and the dream of researchers for many years."

5.2.2 Faster Protocol

As the Internet becomes an ever more vital communications medium for both businesses and consumers, speed becomes an increasingly critical factor. Speed is not only important in terms of rapid data access but also for sharing information between Internet resources.

Soon, Internet-linked systems may be able to transfer data at rates much speedier than is currently possible. That's because a Penn State researcher has developed a faster method for more efficient sharing of widely distributed Internet resources, such as Web services, databases, and high-performance computers. This development has important long-range implications for virtually any business that markets products or services over the Internet.

Jonghun Park, the protocol's developer and an assistant professor in Penn State's School of Information Sciences and Technology, says his new technology speeds the allocation of Internet resources by up to 10 times. "In the near future, the demand for collaborative Internet applications will grow," says Park. "Better coordination will be required to meet that demand, and this protocol provides that."

Park's algorithm enables better coordination of Internet applications in support of large-scale computing. The protocol uses parallel rather than serial methods to process requests. This ability helps provide more efficient resource allocation and also solves the problems of deadlock and livelock—an endless loop in program execution—both of which are caused by multiple concurrent Internet applications competing for Internet resources.

The new protocol also allows Internet applications to choose from among available resources. Existing technology can't support making choices, thereby limiting its utilization. The protocol's other key advantage is that it is decentralized, enabling it to function with its own information. This allows for collaboration across multiple, independent organizations within the Internet's open environment. Existing protocols require communication with other applications, but this is not presently feasible in the open environment of today's Internet.

Internet computing—the integration of widely distributed computational and informational resources into a cohesive network—allows for a broader exchange of information among more users than is possible today. (users include the military, government, and businesses). One example of Internet collaboration is grid computing. Like electricity grids, grid computing harnesses available Internet resources in support of large-scale, scientific computing. Right now, the deployment of such virtual organizations is limited because they require a highly sophisticated method to coordinate resource allocation. Park's decentralized protocol could provide that capability.

Caltech computer scientists have developed a new data transfer protocol for the Internet that is fast enough to download a full-length DVD movie in less than five seconds. The protocol is called FAST, standing for Fast Active queue management Scalable Transmission Control Protocol (TCP). The researchers have achieved a speed of 8,609 Mbps by using 10 simultaneous flows of data over routed paths, the largest aggregate throughput ever accomplished in such a configuration. More importantly, the FAST protocol sustained this speed using standard packet size, stably over an extended period on shared networks in the presence of background traffic, making it adaptable for deployment on the world's high-speed production networks.

The experiment was performed in November 2002 by a team from Caltech and the Stanford Linear Accelerator Center (SLAC), working in partnership with the European Organization for Nuclear Research (CERN), and the organizations DataTAG, StarLight, TeraGrid, Cisco, and Level(3). The FAST protocol was developed in Caltech's Networking Lab, led by Steven Low, associate professor of computer science and electrical engineering. It is based on theoretical work done in collaboration with John Doyle, a professor of control and dynamical systems, electrical engineering, and bioengineering at Caltech, and Fernando Paganini, associate professor of electrical engineering at UCLA. It builds on work from a growing community of theoreticians interested in building a theoretical foundation of the Internet, an effort led by Caltech. Harvey Newman, a professor of physics at Caltech, says the fast protocol "represents a milestone for science, for grid systems, and for the Internet."

"Rapid and reliable data transport, at speeds of 1 to 10 Gbps and 100 Gbps in the future, is a key enabler of the global collaborations in physics and other fields," Newman says. "The ability to extract, transport, analyze, and share many Terabyte-scale data collections is at the heart of the process of search and discovery for new scientific knowledge. The FAST results show that the high degree of transparency and performance of networks, assumed implicitly by Grid systems, can be achieved in practice. In a broader context, the fact that 10 Gbps wavelengths can be used efficiently to transport data at maximum speed end to end will transform the future concepts of the Internet."

Les Cottrell of SLAC, added that progress in speeding up data transfers over long distance are critical to progress in various scientific endeavors. "These include sciences such as high-energy physics and nuclear physics, astronomy, global weather predictions, biology, seismology, and fusion; and industries such as aerospace, medicine, and media distribution. Today, these activities often are forced to share their data using literally truck or plane loads of data," Cottrell says. "Utilizing the network can dramatically reduce the delays and automate today's labor intensive procedures."

The ability to demonstrate efficient high-performance throughput using commercial off-the-shelf hardware and applications, is an important achievement.

With Internet speeds doubling roughly annually, we can expect the performances demonstrated by this collaboration to become commonly available in the next few years; this demonstration is important to set expectations, for planning, and to indicate how to utilize such speeds.

The testbed used in the Caltech/SLAC experiment was the culmination of a multi-year effort, led by Caltech physicist Harvey Newman's group on behalf of the international high energy and nuclear physics (HENP) community, together with CERN, SLAC, Caltech Center for Advanced Computing Research (CACR), and other organizations. It illustrates the difficulty, ingenuity, and importance of organizing and implementing leading-edge global experiments. HENP is one of the principal drivers and codevelopers of global research networks. One unique aspect of the HENP testbed is the close coupling between research and development (R&D) and production, where the protocols and methods implemented in each R&D cycle are targeted, after a relatively short time delay, for widespread deployment across production networks to meet the demanding needs of data-intensive science.

The congestion control algorithm of the present Internet was designed in 1988 when the Internet could barely carry a single uncompressed voice call. Today, this algorithm cannot scale to anticipated future needs, when networks will be to carry millions of uncompressed voice calls on a single path or to support major science experiments that require the on-demand rapid transport of gigabyte to terabyte data sets drawn from multi-petabyte data stores. This protocol problem has prompted several interim remedies, such as the use of nonstandard packet sizes or aggressive algorithms that can monopolize network resources to the detriment of other users. Despite years of effort, these measures have been ineffective or difficult to deploy.

These efforts, however, are necessary steps in our evolution toward ultrascale networks. Sustaining high performance on a global network is extremely challenging and requires concerted advances in both hardware and protocols. Experiments that achieve high throughput either in isolated environments or with interim remedies that by-pass protocol instability, idealized or fragile as they may be, push the state of the art in hardware. The development of robust and practical protocols means that the most advanced hardware will be effectively used to achieve ideal performance in realistic environments.

The FAST team is addressing protocol issues head on to develop a variant of TCP that can scale to a multi-gigabit-per-second regime in practical network conditions. This integrated approach combining theory, implementation, and experiment is what makes the FAST team research unique and what makes fundamental progress possible.

With the use of standard packet size supported throughout today's networks, TCP presently achieves an average throughput of 266 Mbps, averaged over an hour, with a single TCP/IP flow between Sunnyvale near SLAC and CERN in Geneva, over a distance of 10,037 km. This represents an efficiency of just

27 percent. The FAST TCP sustained an average throughput of 925 Mbps and an efficiency of 95 percent, a 3.5-times improvement, under the same experimental condition. With 10 concurrent TCP/IP flows, FAST achieved an unprecedented speed of 8,609 Mbps, at 88 percent efficiency, which is 153,000 times that of today's modem and close to 6,000 times that of the common standard for ADSL (asymmetric digital subscriber line) connections.

The 10-flow experiment set another first in addition to the highest aggregate speed over routed paths. High capacity and large distances together cause performance problems. Different TCP algorithms can be compared using the product of achieved throughput and the distance of transfer, measured in bit-meter-per-second, or bmps. The world record for the current TCP is 10 peta (1 followed by 16 zeros) bmps, using a nonstandard packet size. However, the Caltech/SLAC experiment transferred 21 terabytes over six hours between Baltimore and Sunnyvale using standard packet size, achieving 34 peta bmps. Moreover, data were transferred over shared research networks in the presence of background traffic, suggesting that FAST can be backward compatible with the current protocol. The FAST team has started to work with various groups around the world to explore testing and deployment of FAST TCP in communities that urgently need multi-Gbps networking.

The demonstrations used a 10-Gbps link donated by Level(3) between StarLight (Chicago) and Sunnyvale, as well as the DataTAG 2.5-Gbps link between StarLight and CERN, the Abilene backbone of Internet2, and the TeraGrid facility. The network routers and switches at StarLight and CERN were used together with a GSR 12406 router loaned by Cisco at Sunnyvale, additional Cisco modules loaned at StarLight, and sets of dual Pentium 4 servers each with dual Gigabit Ethernet connections at StarLight, Sunnyvale, CERN, and the SC2002 show floor provided by Caltech, SLAC, and CERN. The project is funded by the National Science Foundation, the Department of Energy, the European Commission, and the Caltech Lee Center for Advanced Networking.

One of the drivers of these developments has been the HENP community, whose explorations at the high-energy frontier are breaking new ground in our understanding of the fundamental interactions, structures, and symmetries that govern the nature of matter and space-time in our universe. The largest HENP projects each encompasses 2,000 physicists from 150 universities and laboratories in more than 30 countries.

Rapid and reliable data transport, at speeds of 1 to 10 Gbps and 100 Gbps in the future, is key to enabling global collaborations in physics and other fields. The ability to analyze and share many terabyte-scale data collections, accessed and transported in minutes, on the fly, rather than over hours or days as is the present practice, is at the heart of the process of search and discovery for new scientific knowledge. Caltech's FAST protocol shows that the high degree of transparency and performance of networks, assumed implicitly by Grid systems, can be achieved in practice.

This will drive scientific discovery and utilize the world's growing bandwidth capacity much more efficiently than has been possible until now.

5.3 GRID COMPUTING

Grid computing enables the virtualization of distributed computing and data resources such as processing, network bandwidth, and storage capacity to create a single system image, giving users and applications seamless access to vast IT capabilities. Just as an Internet user views a unified instance of content via the Web, a grid user essentially sees a single, large virtual computer.

At its core, grid computing is based on an open set of standards and protocols—such as the Open Grid Services Architecture (OGSA)—that enable communication across heterogeneous, geographically dispersed environments. With grid computing, organizations can optimize computing and data resources, pool them for large-capacity workloads, share them across networks, and enable collaboration.

In fact, grid can be seen as the latest and most complete evolution of more familiar developments—such as distributed computing, the Web, peer-to-peer computing, and virtualization technologies. Like the Web, grid computing keeps complexity hidden: multiple users enjoy a single, unified experience. Unlike the Web, which mainly enables communication, grid computing enables full collaboration toward common business goals.

Like peer-to-peer, grid computing allows users to share files. Unlike peer-to-peer, grid computing allows many-to-many sharing—not only files but other resources as well. Like clusters and distributed computing, grids bring computing resources together. Unlike clusters and distributed computing, which need physical proximity and operating homogeneity, grids can be geographically distributed and heterogeneous. Like virtualization technologies, grid computing enables the virtualization of IT resources. Unlike virtualization technologies, which virtualize a single system, grid computing enables the virtualization of vast and disparate IT resources.

5.4 INFOSTRUCTURE

The National Science Foundation (NSF) has awarded $13.5 million over five years to a consortium led by the University of California, San Diego (UCSD) and the University of Illinois at Chicago (UIC). The funds will support design and development of a powerful distributed cyber "infostructure" to support data-intensive scientific research and collaboration. Initial application efforts will be in bioscience and earth sciences research, including environmental, seismic, and remote sensing. It is one of the largest information technology research (ITR) grants awarded since the NSF established the program in 2000.

Dubbed the "OptIPuter"—for optical networking, Internet protocol, and computer storage and processing—the envisioned infostructure will tightly couple computational, storage, and visualization resources over parallel optical networks with the IP communication mechanism. "The opportunity to build and experiment with an OptIPuter has arisen because of major technology changes in the last five years," says principal investigator Larry Smarr, director of the California Institute for Telecommunications and Information Technology [Cal-(IT)2], and Harry E. Gruber Professor of Computer Science and Engineering at UCSD's Jacobs School of Engineering. "Optical bandwidth and storage capacity are growing much faster than processing power, turning the old computing paradigm on its head: we are going from a processor-centric world, to one centered on optical bandwidth, where the networks will be faster than the computational resources they connect."

The OptIPuter project will enable scientists who are generating massive amounts of data to interactively visualize, analyze, and correlate their data from multiple storage sites connected to optical networks. Designing and deploying the OptIPuter for grid-intensive computing will require fundamental inventions, including software and middleware abstractions to deliver unique capabilities in a lambda-rich world. (A "lambda," in networking parlance, is a fully dedicated wavelength of light in an optical network, each already capable of bandwidth speeds from 1 to 10 Gbps.) The researchers in southern California and Chicago will focus on new network-control and traffic-engineering techniques to optimize data transmission, new middleware to bandwidth-match distributed resources, and new collaboration and visualization to enable real-time interaction with high-definition imagery.

UCSD and UIC will lead the research team, in partnership with researchers at Northwestern University, San Diego State University, University of Southern California, and University of California-Irvine [a partner of UCSD in Cal-(IT)2]. Co-principal investigators on the project are UCSD's Mark Ellisman and Philip Papadopoulos of the San Diego Supercomputer Center (SDSC) at UCSD, who will provide expertise and oversight on application drivers, grid and cluster computing, and data management; and UIC's Thomas A. DeFanti and Jason Leigh, who will provide expertise and oversight on networking, visualization, and collaboration technologies. "Think of the OptIPuter as a giant graphics card, connected to a giant disk system, via a system bus that happens to be an extremely high-speed optical network," says DeFanti, a distinguished professor of computer science at UIC and codirector of the university's Electronic Visualization Laboratory. "One of our major design goals is to provide scientists with advanced interactive querying and visualization tools, to enable them to explore massive amounts of previously uncorrelated data in near real time." The OptIPuter project manager will be UIC's Maxine Brown. SDSC will provide facilities and services, including access to the NSF-funded TeraGrid and its 13.6 teraflops of cluster computing power distributed across four sites.

The project's broad multidisciplinary team will also conduct large-scale, application-driven system experiments. These will be carried out in close conjunction with two data-intensive e-science efforts already underway: NSF's EarthScope and the Biomedical Informatics Research Network (BIRN) funded by the National Institutes of Health (NIH). They will provide the application drivers to ensure a useful and usable OptIPuter design. Under co-PI Ellisman, UCSD's National Center for Microscopy and Imaging Research (NCMIR) is driving the BIRN neuroscience application, with an emphasis on neuroimaging. Under the leadership of UCSD's Scripps Institution of Oceanography's deputy director and acting dean John Orcutt, Scripps' Institute of Geophysics and Planetary Physics is leading the EarthScope geoscience effort, including acquisition, processing, and scientific interpretation of satellite-derived remote sensing, near-real-time environmental, and active source data.

The OptIPuter is a "virtual" parallel computer in which the individual "processors" are widely distributed clusters; the "memory" is in the form of large distributed data repositories; "peripherals" are very large scientific instruments, visualization displays, and/or sensor arrays; and the "motherboard" uses standard IP delivered over multiple dedicated lambdas. Use of parallel lambdas will permit so much extra bandwidth that the connection is likely to be uncongested. "Recent cost breakthroughs in networking technology are making it possible to send multiple lambdas down a single piece of customer-owned optical fiber," says co-PI Papadopoulos. "This will increase potential capacity to the point where bandwidth ceases to be the bottleneck in the development of metropolitan-scale grids."

According to Cal-(IT)2's Smarr, grid-intensive applications "will require a large-scale distributed information infrastructure based on petascale computing, exabyte storage, and terabit networks." A petaflop is 1,000-times faster than today's speediest parallel computers, which process one trillion floating-point operations per second (teraflops). An exabyte is a billion gigabytes of storage, and terabit networks will transmit data at one trillion bits per second some 20 million times faster than a dialup 56K Internet connection.

The southern California- and Chicago-based research teams already collaborate on large-scale cluster networking projects and plan to prototype the OptIPuter initially on campus, metropolitan, and state-wide optical fiber networks [including the Corporation for Education Network Initiatives in California's experimental developmental network CalREN-XD in California and the Illinois Wired/Wireless Infrastructure for Research and Education (I-WIRE) in Illinois].

Private companies will also collaborate with university researchers on the project. IBM is providing systems architecture and performance help, and Telcordia Technologies will work closely with the network research teams to contribute its optical networking expertise. "The OptIPuter project has the potential for extraordinary innovations in both computing and networking, and we are pleased to be a part of this team of highly qualified and experienced researchers," says Richard S. Wolff, Vice President of Applied Research at Telcordia. Furthermore, the San Diego Telecom Council, which boasts a membership of 300 telecom companies, has expressed interest in extending OptIPuter links to a variety of public- and private-sector sites in San Diego County.

The project will also fund what is expected to be the country's largest graduate-student program for optical networking research. The OptIPuter will also extend into undergraduate classrooms, with curricula and research opportunities to be developed for UCSD's new Sixth College. Younger students will also be exposed to the OptIPuter, with field-based curricula for Lincoln Elementary School in suburban Chicago and UCSD's Preuss School (a San Diego City charter school for grades 6–12, enrolling low-income, first-generation college-bound students).

The new Computer Science and Engineering (CSE) building at UCSD is equipped with one of the most advanced computer and telecommunications networks anywhere. The NSF awarded a $1.8 million research infrastructure grant over five years to UCSD to outfit the building with a Fast Wired and Wireless Grid (FWGrid). "Experimental computer science requires extensive equipment infrastructure to perform large-scale and leading-edge studies," says Andrew Chien, FWGrid principal investigator and professor of computer science and engineering in the Jacobs School of Engineering. "With the FWGrid, our new building will represent a microcosm of what Grid computing will look like five years into the future."

FWGrid's high-speed wireless, wired, computing, and data capabilities are distributed throughout the building. The research infrastructure contains teraflops of computing power, terabytes of memory, and petabytes of storage. Researchers can also access and exchange data at astonishingly high speeds. "Untethered" wireless communication will happen at speeds as high as 1 Gbps, and wired communication will top 100 Gbps. "Those speeds and computing resources will enable innovative next-generation systems and applications," says Chien, who notes that Cal-IT2 is also involved in the project. "The faster communication will enable radical new ways to distribute applications, and give us the opportunity to manipulate and process terabytes of data as easily as we handle megabytes today."

Three other members of Jacobs School's computer science faculty will participate in the FWGrid project. David Kriegman leads the graphics and image processing efforts, whereas Joseph Pasquale and Stefan Savage are responsible, respectively, for the efforts in distributed middleware and network measurement.

Key aspects of this infrastructure include mobile image/video capture and display devices, high-bandwidth wireless to link the mobile devices to the rest of the network, "rich" wired networks of 10–100 Gbps to move and aggregate data and computation without limit, and distributed clusters with large processing (teraflops) and data (tens of terabytes) capabilities (to power the infrastructure). "We see FWGrid as three concentric circles," explained Chien. "At the center will be super-high-bandwidth networks, large compute servers, and data storage centers. The middle circle includes wired high bandwidth, desktop compute platforms, and fixed cameras. And at the mobile periphery will be wireless high bandwidth, mobile devices with large computing and data capabilities, and arrays of small devices such as PDAs, cell phones, and sensors."

Because FWGrid will be a living laboratory, the researchers will gain access to real users and actual workloads. "This new infrastructure will have a deep impact on undergraduate and graduate education," says CSE chair Ramamohan Paturi. "It will support experimental research, especially cross-disciplinary research. It will also provide an opportunity for our undergraduates to develop experimental applications." Research areas to be supported by FWGrid include low-level network measurement and analysis, grid middleware and modeling, application-oriented middleware, new distributed application architectures, and higher-level applications using rich image and video, e.g., enabling mobile users to capture and display rich, three-dimensional information in a fashion that interleaves digital information with reality.

5.4.1 Intelligent Agents

Intelligent agents—network programs that learn over time to understand its user's likes and dislikes, needs, and desires—will help people find the right products and services. Agents will also help users quickly locate important information, ranging from a friend's phone number to the day's closing stock prices. "I, as a user, won't even think about it anymore," says Bell Labs' Sweldens. "I will have the perception that the network knows everything it needs to know and it has the right information available at the right time."

Researchers at Carnegie Mellon University's School of Computer Science (SCS) have embarked on a five-year project to develop a software-based cognitive personal assistant that will help people cut through the clutter caused by the arrival of advanced communications technologies.

The Reflective Agents with Distributed Adaptive Reasoning (RADAR) project aims to help people schedule meetings, allocate resources, create coherent reports from snippets of information, and manage e-mail by grouping related messages, flagging high-priority requests and automatically proposing answers to routine messages. The ultimate goal is to develop a system that can both save time for its user and improve the quality of decisions.

RADAR will be designed to handle some routine tasks by itself, to ask for confirmation on others, and to produce suggestions and drafts that its user can modify as needed. Over time, the researchers hope the system will be able to learn when and how often to interrupt its busy user with questions and suggestions. To accomplish all of this, the RADAR research team will draw techniques from a variety of fields, including machine learning, human-computer interaction, natural-language processing, optimization, knowledge representation, flexible planning, and behavioral studies of human managers.

The project will initially focus on four tasks: e-mail, scheduling, webmastering, and space planning. "With each task, we'll run experiments to see how well people do by themselves and make comparisons," says Daniel P. Siewiorek, director of Carnegie Mellon's Human-Computer Interaction Institute. "We will also look at people, plus a human assistant, and compare that to the software agent." Initial software development goals include the creation of a shared knowledge base, a module that decides when to interrupt the user with questions, and a module that extracts information, such as meeting requests, from e-mail messages.

"The key scientific challenge in this work is to endow RADAR with enough flexibility and general knowledge to handle tasks of this nature," says Scott Fahlman, one of project's principal researchers. "Like any good assistant, RADAR must understand its human master's activities and preferences and how they change over time. RADAR must respond to specific instructions, such as 'Notify me as soon as the new budget numbers arrive by e-mail,' without the need for reprogramming." Yet Fahlman notes that the system also must be able to learn by interacting with its master to see how he or she reacts to various events. "It must know when to interrupt its master with a question and when to defer," he says.

The project has received an initial $7 million in funding from the Defense Advanced Research Projects Agency (DARPA), which is interested in the technology's potential to streamline military communications and workloads.

5.4.2 Next-Generation Agent

A new intelligent agent that works through users' mobile phones to organize business and social schedules has been developed by scientists at a university in the United Kingdom.

Artificial intelligence software allows the agent to determine users' preferences and to use the Web to plan business and social events, such as travel itineraries and visits to restaurants and theatres. "I see the artificial agent as a butler-type character," says Nick Jennings, professor of computer science at the University of Southampton's electronics and computer department. "The first day that the 'butler' comes to work, he will be very polite, as he does not know much about me. But as we begin to work together, he will become better acquainted with my preferences and will make decisions without having to consult me. The degree of autonomy I allow him is entirely up to me."

Jennings believes that his research team's agent will work well with existing 3G mobile networks. It will reduce the need for business travelers to carry laptop computers, since they will be able to do their computing through their phone.

Jennings and his team are among the U.K.'s leading artificial intelligence researchers. Earlier this year they won the ACM Autonomous Research Award in recognition of their research in the area of autonomous agents. Last year, Jenning's team developed an agent that functioned as a virtual travel agent, producing the best possible vacations based on clients' preferences, including budgets, itineraries, and cultural visits. All of the travel package's component's had to be purchased from a series of online auctions. "Here we had a scenario where artificial agents outperformed humans as they assimilated information much more quickly than any human could possibly operate," says Jennings. "The world is getting more complicated, so the more support we have with planning and taking decisions, the better we can function."

Although Jennings' 3G intelligent agent shows much promise, it's unlikely that it will be able to fully meet the varied information needs of mobile phone users. "It's very difficult to second guess what people have on their minds," says Alex Linden, a vice president of research at Gartner, a technology research firm based in Stamford, Connecticut. Linden notes that mobile phone users' requirements change quickly, depending on their mood, physical location, and personal or work situation, making it difficult for an agent to keep pace. "I may be calling someone for myself or on behalf of my boss, my wife, or my kids," says Linden. "The system would have to switch radically between different contexts."

Linden notes that many university and corporate laboratories are working on agent research. "But it's going to be a long time before something useful comes out of that," he says.

5.5 TELE-LEARNING OPENS HORIZONS

Tele-learning may actually be better than in-person instruction, since online classes can more easily expose students to new cultures and concepts, says an Ohio State University researcher.

Courses taught on the Web allow students to interact with people from around the world and to learn new perspectives that they could never experience in a typical classroom, says Merry Merryfield, professor of social studies and global education at Ohio State University. She notes that online classes also allow students to tackle more controversial subjects, ensure that all students participate equally, and offer the opportunity for more thoughtful and in-depth discussions.

Merryfield teaches graduate-level online classes on global education to teachers in the United States and around the world. In a recent study, she examined the online interaction of 92 U.S. teachers who took her courses and 22 cultural consultants—educators from other countries that Merryfield hired to provide the American teachers with international perspectives. The results, she says, showed the value of online classes in global education. "Online technologies provide opportunities for teachers to experience a more global community than is possible face-to-face," says Merryfield. "In a course I taught last summer, I had 65 people from 18 states and 12 countries. This diversity affected the course and the content in many ways and greatly helped the learning process."

Merryfield also found that Web-based interaction allows discussion of sensitive and controversial topics that would be difficult in face-to-face settings. As a result, students can tackle cultural and political issues—such as those involving terrorism and the war with Iraq—that they might be reluctant to do in a classroom. One student told Merryfield that "online discussions are like a veil that protects me" and allowed her to "feel safe enough to ask the hard questions" of other students in her class. The student noted, "People respond to text instead of a person's physical presence, personality, accent, or body language."

Online classes have another important advantage: they allow all students to participate equally. In traditional classes, it is found that discussion are generally dominated by a few students. In her Web-based classes, Merryfield established rules that set minimum and maximum numbers of messages each person posts. "There is no possibility of a few people monopolizing a discussion, nor is anyone left out," Merryfield says.

One of the main ways students communicate in the online class is through threaded discussion—an interactive discussion in which a person posts a message, people respond to it, and then people can respond to those responses. Threaded discussions can be much more meaningful and in-depth than traditional classroom oral discussions, says Merryfield. "Discussions take place over several days, so people have time to look up references and share resources," she says. "They have time to think, analyze, and synthesize ideas. I have been amazed at how these threaded discussions increase both the depth of content and equity in participation."

Overall, online courses can provide significant advantages for teaching global education, Merryfield says. "Online technologies are the perfect tools for social studies and global education, as these fields focus on learning about the world and its peoples." "This provides opportunities for teachers to experience a more global community than is possible face to face." However, she adds, "All of us do need opportunities for face to face experiential learning with people of diverse cultures."

5.6   A NEW APPROACH TO VIRUS SCANNING

A new technology, developed by a computer scientist at Washington University in St. Louis, Missouri, aims to stop malicious software, including viruses and worms, long before it has a chance to reach home or office computers.

John Lockwood, an assistant professor of computer science at Washington University, and the graduate students that work in his research laboratory have developed a hardware platform called the field-programmable port extender

(FPX). This system scans for malicious codes transmitted over a network and filters out unwanted data.

The FPX is an open platform that augments a network with reprogrammable hardware. It enables new data-processing hardware to be rapidly developed, prototyped, and deployed over the Internet. "The FPX uses several patented technologies in order to scan for the signatures of 'malware' quickly," says Lockwood. "The FPX can scan each and every byte of every data packet transmitted through a network at a rate of 2.4 billion bits per second. In other words, the FPX could scan every word in the entire works of Shakespeare in about 1/60th of a second."

Viruses spread when a computer user downloads unsafe software, opens a malicious attachment, or exchanges infected computer programs over a network. An Internet worm spreads over the network automatically when malicious software exploits one or more vulnerabilities in an operating system, a Web server, a database application, or an e-mail exchange system. Existing firewalls do little to protect against such attacks. Once a few systems are compromised, they proceed to infect other machines, which in turn quickly spread throughout a network.

Recent attacks by codes such as Nimba, Code Red, Slammer, SoBigF, and MSBlast have infected computers globally, clogged large computer networks, and degraded corporate productivity. Once an outbreak hits, it can take anywhere from weeks to months to cleanse all of the computers located on an afflicted network.

"As is the case with the spread of a contagious disease like SARS, the number of infected computers will grow exponentially unless contained," says Lockwood. "The speed of today's computers and vast reach of the Internet, however, make a computer virus or Internet worm spread much faster than human diseases. In the case of SoBigF, over 1 million computers were infected within the first 24 hours and over 200 million computers were infected within a week."

Most Internet worms and viruses aren't detected until after they reach a user's PC. As a result, it's difficult for enterprises to maintain network-wide security. "Placing the burden of detection on the end user isn't efficient or trustworthy because individuals tend to ignore warnings about installing new protection software and the latest security updates," notes Lockwood. "New vulnerabilities are discovered daily, but not all users take the time to download new patches the moment they are posted. It can take weeks for an IT department to eradicate old versions of vulnerable software running on end-system computers."

The FPX's high speed is possible because its logic is implemented as field programmable gate array (FPGA) circuits. The circuits are used in parallel to scan and to filter Internet traffic for worms and viruses. Lockwood's group has developed and implemented FPGA circuits that process the Internet protocol (IP) packets directly in hardware. This group has also developed several circuits that rapidly scan streams of data for strings or regular expressions in order to find the signatures of malware carried within the payload of Internet packets. "On the FPX, the reconfigurable hardware can be dynamically reconfigured over the network to search for new attack patterns," says Lockwood. "Should a new Internet worm or virus be detected, multiple FPX devices can be immediately programmed to search for their signatures. Each FPX device then filters traffic passing over the network so that it can immediately quarantine a virus or Internet worms within subnetworks [subnets]. By just installing a few such devices between subnets, a single device can protect thousands of users. By installing multiple devices at key locations throughout a network, large networks can be protected."

Global Velocity, a St. Louis-based firm, has begun building commercial systems that use FPX technology. The company is working with corporations, universities, and the government to install systems in both local area and wide area networks. The device self-integrates into existing Gigabit Ethernet or Asynchronous Transfer Mode (ATM) networks. The FPX fits within a rack-mounted chassis that can be installed in any network closet. When a virus or worm is detected, the system can either silently drop the malicious traffic or generate a pop-up message on an end-user's computer. An administrator uses a simple Web-based interface to control and configure the system.

5.7 PUTTING A LID ON SPAM

It's no secret that spam is spiraling out of control and that anti-spam filters do a generally poor job of blocking unwanted e-mail.

In an effort to give e-mail users more control over spam, professors Richard Lipton and Wenke Lee of Georgia Tech's Information Security Center have created a new application that offers a different approach to reducing spam. "What does the spammer want the e-mail user to do?" asks Lee. "Usually, the spammer wants the recipient to click on a link to a Web address to find out more about the product or service and buy it online." In thinking about the problem from the spammer's point of view, Lee and Lipton realized that most spam e-mail contains a URL or Web address for potential customers to click. So they created a filter application based on looking for unwanted URL addresses in e-mails. "This approach and application is elegant and incredibly computer cheap and fast," says Lipton. "It seems to work better than the existing commercial products and the end user can customize it easily."

Lee developed a working prototype over the past year, and the two researchers have recently run the software on several computers. The developers are very pleased with the results. The software directs all e-mails that don't contain an embedded Web address into the e-mail client's inbox. The end user can also create "white lists"—the opposite of black lists—of URLs that are acceptable, such as favorite news sites or online retailers. A "wild card"

category lets users tell the system to allow e-mails containing specified parameters, such as messages with an ".edu." extension.

The application also includes a "black list" feature that allows users to easily
add URLs from unwanted e-marketers and others. The blacklisted e-mails are
automatically delivered to a "Spam Can" that can be periodically checked to
make sure no wanted e-mails were accidentally trashed. "We've had very few
false positives," says Lipton. "It's important that the system not accidentally
remove legitimate e-mail."

Lipton and Lee have obtained a provisional patent on their new anti-spam
tool. They plan to refine the application by adding several more customizable
filtering features, finalize the patent, and write a paper about their project. The
researchers hope to eventually license the as-yet unnamed application for
widespread business and consumer use.

5.8 THE MEANING BEHIND MESSAGES

Speak to someone on the telephone and you can tell by the way a person
sounds—the tone of voice, inflections, and other nonverbal cues—exactly what
the speaker means. These cues are missing from text-based e-mail and instant
messages, leading users to often wonder what a particular message really
means and whether the sender is being truthful.

Researchers at the University of Central Florida are now looking into how
people form opinions about others through e-mail and instant messages.
Michael Rabby, a professor at the university's Nicholson School of Communication, along with graduate student Amanda Coho, wants to find out what
messages and phrases are most likely to make people believe that someone
who is e-mailing them is, among other things, trustworthy, eager to help others,
and willing to admit to making mistakes. "If I want to show you that I always
go out of my way to help people in trouble, what messages would I send to
convey that?" asks Rabby. "In business, if I want to show that I'm compas-
sionate, how would I do that?"

Early results of Rabby's and Coho's research show that people generally
develop favorable opinions of someone with whom they're communicating
only via e-mail or instant messages. Rabby says it's easier to make a favorable
impression in part because people can choose to present only positive information about themselves.

Rabby recently divided students in his Communication Technology and
Change classes into pairs and asked them to exchange five e-mail messages
with their partners. In almost all of the cases, the students did not know each
other and formed first impressions based on the e-mail messages and, sometimes, instant messages. After they exchanged the e-mail messages, the students filled out surveys rating themselves and their partners in a variety of
categories, such as whether they're likely to go out of their way to help

someone in trouble and whether they practice what they preach. In most cases, the students gave their partners higher ratings than themselves.

The students also reexamined e-mail messages written by their partners and highlighted specific phrases or sections that caused them to form opinions. In some cases, direct statements such as "I am independent" led to conclusions, whereas other opinions were based on more anecdotal evidence. One student wrote about how he quit a "crooked" sales job to keep his integrity, and another student mentioned that she soon would be helping her pregnant roommate care for a baby that was due in a few weeks.

"E-mail communication has impacted so much of our lives, but it's still a pretty understudied area," says Rabby. "We're now looking more closely at the messages that impact people the most.

5.9   INTERNET SIMULATOR

Researchers at the Georgia Institute of Technology have created the fastest detailed computer simulations of computer networks ever constructed—simulating networks containing more than 5 million network elements. This work will lead to improved speed, reliability, and security of future networks such as the Internet, according to Professor Richard Fujimoto, lead principal investigator of the DARPA-funded project (Defense Advanced Research Projects Agency).

These "packet-level simulations" model individual data packets as they travel through a computer network. Downloading a Web page to an individual's home computer or sending an e-mail message typically involves transmitting several packets through the Internet. Packet-level simulations provide a detailed, accurate representation of network behavior (e.g., congestion) but are very time consuming to complete.

Engineers and scientists routinely use such simulations to design and analyze new networks and to understand phenomena such as Denial of Service attacks that have plagued the Internet in recent years. Because of the time required to complete the simulation computations, most studies today are limited to modeling a few hundred network components such as routers, servers, and end-user computers.

"The end goal of research on network modeling and simulation is to create a more reliable and higher-performance Internet," says Fujimoto. "Our team has created a computer simulation that is two to three orders of magnitude faster than simulators commonly used by networking researchers today. This finding offers new capabilities for engineers and scientists to study large-scale computer networks in the laboratory to find solutions to Internet and network problems that were not possible before."

The Georgia Tech researchers have demonstrated the ability to simulate network traffic from over 1 million Web browsers in near real time. This feat means that the simulators could model one minute of such large-scale network operations in only a few minutes of clock time.

Using the high-performance computers at the Pittsburgh Supercomputing Center, the Georgia Tech simulators used as many as 1,534 processors to simultaneously work on the simulation computation, enabling them to model more than 106 million packet transmissions in one second of clock time—two to three orders of magnitude faster than simulators commonly used today. In comparison, the next closest packet-level simulations of which the research team is aware have simulated only a few million packet transmissions per second.

5.10 UNTANGLING TANGLED NETS

New software developed by Ipsum Networks, a start-up cofounded by a University of Pennsylvania engineering professor, shows promise in detecting hard-to-spot bottlenecks in computer networks. The first version of this software, known as Route Dynamics, is available to companies and other users that transmit data via decentralized IP networks.

"IP networks have gained popularity because they don't rely on a central computer and are therefore more resistant to attacks and failure, but this complex architecture also makes them much harder to monitor and repair," says Roch Guerin, professor of electrical and systems engineering at Penn and CEO of Ipsum. "Managing an IP network is now a labor-intensive art rather than an automated science. Making matters worse, corporations can lose literally millions of dollars for every second their IP network is down."

IP networks work by dividing data into packets, which are addressed and then transmitted by way of a series of routers. A router detects a packet's ultimate address and communicates with other routers before sending the packet to another machine that it believes is closer to the packet's final destination. Route Dynamics monitors communications between routers as well as communications between entire networks, a level of surveillance not attainable with existing programs, which measure only network speed or simply monitor devices. The new software may help assuage corporations' concerns about moving business-critical functions onto IP networks.

"Ultimately, an IP network is only as good as the communication between its routers," says Guerin, who founded Ipsum with one-time IBM colleague Raju Rajan, now Ipsum's chief technology officer. "When routers share inaccurate information, it can slow or freeze a network; such performance difficulties are generally the first sign of trouble. But ideally you'd like to catch the problem before network performance is compromised."

Because it's nearly impossible even for skilled network administrators to spot less-than-optimal communication between routers, such problems can take a significant amount of time and money to solve. Performance sometimes suffers for extended periods as computer professionals attempt to identify the problem, making organizations increasingly interested in automating monitoring of IP networks.

In addition to monitoring existing networks, Route Dynamics can also predict the performance of a network. If a user wants to determine how an added piece of equipment will affect a network's performance, Route Dynamics can perform simulations.

Chapter **6**

Something in the Air—Radio and Location Technologies

Over the past century, radio has experienced periodic technological revolutions. In the 1920s, amplitude modulation (AM) voice technology began replacing spark gap telegraphy. In the 1950s, frequency modulation (FM) started its gradual yet relentless takeover of AM technology. Now digital radio, which replaces analog modulation technologies with data streams, and software-defined radios, which allow a device to be configured into any type of radio its user desires, are the next great leaps forward in radio technology.

6.1 DIGITAL RADIO

Digital radio already surrounds us. Most mobile phones are already digital. And as subscribers to the XM Radio and Sirius services will readily attest, digital broadcast radio is already here (following digital television, which debuted in the 1990s on some satellite and cable TV services). Over the next several years, digital radio will grow rapidly and will eventually replace today's AM-FM receivers.

The worldwide digital radio market, both satellite and terrestrial, will grow to over 19 million unit shipments in 2007, according to statistics compiled by Scottsdale, Arizona-based In-Stat/MDR. The high-tech market research firm believes that new content—stations that only exist in digital—and data ser-

Telecosmos: The Next Great Telecom Revolution, edited by John Edwards
ISBN 0-471-65533-3 Copyright © 2005 by John Wiley & Sons, Inc.

vices will drive consumer demand for these radios. These factors are already at work in the digital satellite radio arena in the United States and the digital terrestrial market in the United Kingdom. "The conversion from analog radio to digital has been a long, slow process that will take many more years," says Michelle Abraham, a senior analyst with In-Stat/MDR. "When the first digital broadcasts became available in Europe, receivers were too expensive for the mass market. Over five years later, receiver prices have come down, but many countries are still trialing digital broadcasts, waiting for the regulatory framework to be in place and digital coverage to expand."

Satellite radio has been successful in the United States, and other countries are hoping to duplicate that success. In South Korea and Japan, providers want to deliver not only audio streams but video streams as well. Several hundred million analog radios are sold worldwide each year, in the form of stereo receivers, CD boom boxes, portable devices, alarm clocks, and car stereo systems. Reductions in the cost of digital tuners will convert the more expensive of the analog radios to digital by the end of 2007.

Even venerable U.S. AM broadcast radio, where the whole industry started some 80 years ago, is going digital. HD Radio broadcasting technology, developed by Columbia, Maryland-based iBiquity, promises to transform the listening experience of the 96 percent of Americans who listen to AM and FM radio on a weekly basis. The commercial launch of HD Radio will enable the nation's 12,000 radio stations to begin the move from analog broadcasting to higher-quality digital broadcasting, bringing improved audio quality and an array of wireless data services to consumers. HD Radio technology has been tested under experimental licenses across the country, including KLUC-FM and KSFN-AM in Las Vegas, KDFC-FM and KSFN-AM in San Francisco, WILC-AM in Washington DC, WGRV-FM and WWJ-AM in Detroit, and WNEW-FM in New York.

6.2 SOFTWARE-DEFINED RADIO

Standards, such as GSM, GSM, and CDMA, define the way mobile phones are designed and used. But what if a mobile phone or radio could be instantly adapted to accommodate any standard, simply by loading in various free programs? Such a product—a software-defined radio (SDR)—would not only lead to easy, pain-free compatibility but would revolutionize the wireless telecom industry.

Proponents claim that SDR technology represents the future of wireless communications. As a result, the days of custom ASICs and radio hardware may be numbered, ushering in a new era where upgrades and reconfigurations of wireless equipment require only a new software load. SDR has the potential to open new business opportunities for carriers and cellular providers as well as the entire handset market. However, none of these opportunities is expected to happen overnight. Several key development milestones, including the creation of industry-wide standards, must be reached to move SDR technologies into the mainstream market.

SDR is already used in some base station products, as well as in military and aerospace equipment. Although the technology has yet to make major inroads into any consumer-oriented markets, the SDR Forum trade group projects that by 2005 SDR will have been adopted by many telecom vendors as their core platform.

Another potential route to SDR is being pioneered by the open source-based GNU Radio project. Eric Blossom, the project's leader, believes that SDR makes a lot of sense from an open-source standpoint, since it would make phones and radios infinitely compatible and flexible. An SDR radio could, for example, let users simultaneously listen to an FM music station, monitor a maritime distress frequency, and upload data to an amateur radio satellite. Developers could also create a "cognitive radio"—one that seeks out unused radio frequencies for transmissions. "That could go a long way toward solving the current spectrum shortage," says Blossom.

The GNU Radio project also wants to throw a virtual monkey wrench into the efforts of big technology and entertainment companies to dictate how, and on what platforms, content can play. By creating a user-modifiable radio, Blossom and his colleagues are staking a claim for consumer control over those platforms. "The broadcast industry has a business plan that fundamentally hasn't changed since 1920," he says. "I don't see any constitutional guarantee that some previous business plan still has to be viable."

Work on the GNU Radio project's first design—a PC-based FM receiver—is complete, with an HDTV transmitter and receiver in the works. Yet developers still must overcome power consumption issues and other hurdles before a software-defined radio becomes practical. Blossom estimates that it will take about five years before SDR hits the mainstream.

6.3 ULTRAWIDEBAND RADIO

Today's relatively poor-quality wireless local area networks (WLANs) could soon be replaced by a far more sophisticated technology. Ultrawideband (UWB) promises high-bandwidth, noninterfering and secure communications for a wide array of consumer, business, and military wireless devices.

A team of Virginia Tech researchers is attempting to take UWB to a new level. Michael Buehrer and colleagues William Davis, Ahmad Safaai-Jazi, and Dennis Sweeney in the school's mobile and portable radio research group are studying how UWB pulses are propagated and how the pulses can be recognized by potential receivers.

A UWB transmission uses ultrashort pulses that distribute power over a wide portion of the radio frequency spectrum. Because power density is dispersed widely, UWB transmissions ideally won't interfere with the signals on narrow-band frequencies, such as AM or FM radio or mobile phone signals.

In fact, UWB transmissions pose so little threat of interference with licensed frequencies that the FCC now allows companies to operate UWB technology within the 3-GHz to 10-GHz range without obtaining radio spectrum licenses.

The bandwidth of UWB signals is so wide that signal energy is available for use at both high and low frequencies. "The low-frequency content of UWB devices can penetrate solid structures," says Buehrer, an assistant professor of electrical and computer engineering and the project's principal investigator. That would make UWB highly useful for transmitting signals through buildings and other manmade and natural obstacles. "Additionally, the high-frequency content can detect the details of objects," notes Buehrer. This capability, combined with the technology's low power, makes UWB radar an excellent surveillance tool.

UWB also has the potential to become a significant military communications medium. "Because of the low level of energy in UWB signals, a military unit using the technology could communicate without a nearby enemy even perceiving that transmissions are taking place," says Buehrer.

UWB also has many commercial applications. For example, most home wireless devices, such as television remote controls, are limited in the amount of data they can send and receive. UWB signals can achieve significantly higher data rates. As a result, there's a potential for UWB wireless home computer networks, wireless camera to computer downloads, and wireless connections to thin-screen wall-mounted televisions.

In the project's first phase, Buehrer and his colleagues will develop models to show the characteristics of UWB-transmitted pulses and how the pulses will look to receivers. "We'll discover what receivers see when they encounter UWB signals," Buehrer says. The research team hopes to continue the project into a second phase, during which they would use the models developed in the first phase to design UWB receivers. The project is funded by a $750,000 grant from the Defense Advanced Research Projects Agency (DARPA).

Buehrer believes that the FCC will continue to allow UWB devices to operate without licenses, which should help the technology proliferate. "UWB already has a long history," he notes. "The technology has been used in radar devices for some time. Actually, it's been around since Marconi transmitted the first telegraph signals."

6.4 ASSET TRACKING

The Telecosmos promises to give people and businesses unprecedented control over physical assets. In the years ahead, wireless sensors will help organizations track and monitor everything from vending machines to roadways.

Radio Frequency Identification (RFID) has existed for at least a decade, yet the technology has never lived up to its proponents' expectations. "It's always something that's 'the next big thing,'" says Jeff Woods, a senior analyst at Stamford, Connecticut-based Gartner. Like others who follow the industry, Woods believes that RFID's acceptance has been hampered by a number of factors, including high costs, a lack of standards, and global radio frequency differences that sometimes prevent businesses from shipping RFID-tagged objects between countries.

Two separate mandates for 2005, set by Wal-Mart and the U.S. Department of Defense (DoD)—both requiring suppliers to embrace RFID—have pushed the technology into the public eye. As a result, RFID is rapidly moving from a company science experiment to boardroom priority, with a focus on improving enterprise-wide operations. Manufacturers and the suppliers to Wal-Mart and the DoD are diving into an increasingly busy RFID market already brimming with developing standards, large company entrants, start-up software developers, and numerous systems integrators. Despite some recognizable large company names, success is still to be determined, says Erik Michielsen, a senior analyst with technology research ABI, based in Oyster Bay, New York.

Texas Instruments, Symbol Technologies, NCR, Philips, and Sun Microsystems are only some of the big-name companies that have entered the world of RFID. Some recognizable names have entered the RFID fray as systems integrators, namely, IBM, Accenture, BearingPoint, Unisys, RedPrairie, and Manhattan Associates. Process questions abound, such as where to store the data, what data should be stored, how to secure and maintain data, and what is the optimal method to integrate data with existing business solutions. Some integrators, such as SAP, are developing enterprise-level RFID patches for customers. There are others, known as warehouse management systems companies, which include Manhattan Associates, RedPrairie, and Provia. Long-time DoD integration partners such as Unisys, Lockheed Martin, and Accenture are stepping up government-based RFID efforts.

"Due to the time constraints and the still-developing standards, prior relationships will drive RFID integration contracts even more than with previous rollouts, such as ERP or supply chain management systems," notes Michielsen. "This is not necessarily good for the RFID business, as the process discourages competition and rewards relationships over capabilities. The upside is that established relationships will better enable scalable, successful solutions due to better understanding of environment, staff, and business goals."

Another complex issue is that RFID is new and there have been few full-scale projects to date, especially for supply chain solutions. Although integrators such as SCS, Unisys, and Lockheed Martin have extensive, long-term relations with the DoD, they do not have extensive experience with passive, UHF RFID tags. The leading supplier lists for Wal-Mart and the DoD are long, and integration solutions must conform more than differentiate if these projects are going to roll out to specification and on time.

Still, many of RFID's shortcomings are gradually being resolved as the industry's vendors join together to make the technology more attractive to businesses. RFID standards covering agriculture, vehicle management, postal items, and freight containers are at various stages of maturity. Industry observers are hoping that a basic support framework allowing interoperability between vendors' products will take shape within the next couple of years. Costs are gradually coming down as the technology matures. Frequency conflicts are also becoming less of an issue, as vendors and government agencies work together to smooth out global differences. As a result, although an RFID boom isn't in the wings, steady growth appears likely.

Businesses have much to gain by adopting RFID. The technology provides key information more efficiently than bar codes in a variety of environments (even in hurricanes and blizzards) with little or no human intervention. RFID tags can also contain more information than bar codes, making it possible to retrieve information about an asset's type, configuration, version, location, history of location and maintenance, and other facts. The added speed and rich information provided by RFID can lead to significant savings. "Early implementations have shown a 3 percent to 5 percent reduction in supply chain costs and 2 percent to 7 percent increases in revenue from inventory visibility," says Peter Abell, director of retail research for Boston-based AMR Research.

6.4.1 RFID Components

An RFID system consists of two components: tags and readers. Tags (also known as transponders) incorporate a chip and an antenna. Active tags, which include a battery, can transmit hundreds of feet and cost upward of $5. Passive tags are smaller, require no battery, and usually have a range of only a few feet. Thanks to their simplicity, they generally cost less than a dollar.

Readers (sometimes called interrogators) communicate with tags to retrieve and, sometimes, write information to the tag. Readers are designed to work with a specific type of tag in one of the four RFID frequency ranges: 125 to 134 kHz, 13.553 to 13.567 MHz, 400 to 1 GHz, and 2.3 to 2.48 GHz. The reader also relays information into a database and other parts of an organization's IT infrastructure.

Despite its many variations, RFID is a fundamentally simple technology. What isn't so simple, and what has contributed to RFID's slow progression into the mainstream, is its need to mesh with existing business systems and practices. Databases, networks, employee job duties—even warehouse layouts and production lines—must all be tweaked or entirely redesigned to accommodate RFID. "It really changes many business processes throughout the organization," says Gartner's Woods.

Yet RFID can also provide the rationale for a profitable business-line restructuring. Carlsberg-Tetley Brewing, for example, identified RFID as an opportunity to outsource the management of its beer kegs. "It will put the complexities and the rigor of content management into the hands of a better provider," says David Dixon, business solutions executive for the Northampton, England-based beer maker.

Carlsberg-Tetley, one of the United Kingdom's largest brewers, recently sold more than 1 million of its containers (kegs) to Trenstar, a Denver-based asset management company. Under this arrangement, the brewer pays for use of the containers on a "per fill" basis, whereas Trenstar retains legal possession of them. In addition, Trenstar put RFID tags on each container and installed fixed readers alongside conveyors inside Carlsberg-Tetley's breweries. Delivery trucks are also equipped with readers that scan the kegs as before and after delivery.

The arrangement is designed to allow Carlsberg-Tetley to improve its return on capital by removing the containers from its balance sheet. The RFID technology, on the other hand, should let Trenstar cut the losses Carlsberg-Tetley was experiencing from lost and stolen kegs. "That's the result of the need to attach tags to over 1 million containers," Dixon notes.

6.4.2 Tag and Read

As RFID evolves and the prices fall, an ever-wider array of objects will be tracked. Many observers also expect RFID to eventually find a home inside a variety of everyday business and consumer products. "RFID is actually already deployed in many retail environments. People just don't think about it that way," says Woods. For many years, in-store theft prevention systems have relied on RFID-tagged merchandise to snare shoplifters. More than 6 million consumers also carry RFID tags on their key chains in the form of Exxon Mobil Speedpass tokens. The device, when waved in front of a gas pump-mounted reader, sends an identification code that allows the merchant to deduct the purchase amount from a linked credit card or checking account. "It's a great application," says Joe Giordano, vice president of Speedpass network business and product development at Exxon Mobil in Dallas. "I think it could benefit any retailer, particularly retailers who have convenience-type transactions."

Tiny, cheap tags will allow the efficient tracking of even the smallest items, such as overnight letters and packages. An RFID tag attached to a letter would not only tell a shipper the package's current location but also where it's been and where it's scheduled to go. "Pieces of mail will probably wait until [tag prices] get down to one or two cents," says AMR's Abell. Miniature tags— perhaps in the form of an implantable chip—will also allow pet owners to affordably and conveniently track the movements of Fido and Fluffy. "[Wild] animals have been tracked with RFID for a long time," says Abell. "They even put them on hummingbirds."

Then there's the potential for people tracking. Sporting event and concert tickets could incorporate tags that allow event organizers to sidetrack counterfeiting, achieve improved crowd flow management, and ensure that people sit in their assigned seats. Likewise, RFID could help parents track their kids' movements around an amusement park. At Hyland Hills Waterworld, a water-park in Federal Heights, Colorado, RFID wristbands have taken the place of money and credit cards. "People don't carry their wallets or purses when they're in the pool," says Bob Owens the park's assistant manager. "We needed a way to allow people to spend money easily while in the park."

The wristbands work like a wearable debit card, allowing people to spend money simply by waving their hands past readers located at snack bars, gift shops, and other park venues. At the end of the day, the wristband is either thrown away or saved as a souvenir. Although more expensive than barcode imprinted wristbands, the RFID devices aren't vulnerable to damage caused by pool chemicals, the sun, or stretching. "The fact that they're disposable means that that we don't have to worry about the band's long-term physical integrity," says Owens.

More ominously, authoritarian governments could use implantable tags to track people and create lists of places they've visited. "There is a dark side to this technology," says AMR Research's Abell.

6.4.3 RFID in Retail

To find the "store of the future," you'll have to travel to Rheinberg, Germany. That's where METRO Group, the world's fifth largest retailer, has created a convenience store that's designed to serve as a real world test bed for a variety of advanced retailing technologies.

The Extra Future Store, as the outlet is formally known, aims to showcase promising systems that can benefit both shoppers and retailers. Very little inside the store isn't touched by some type of technology. Cases, shipping pallets, shopping carts, and individual products are all tagged with RFID devices that allow everything from sales tracking to automatic inventory replenishment to preventing congestion at the checkout line. Consumers can also take advantage of an intelligent scale that automatically identifies and weighs fruit and vegetables as well as an RFID-based self-checkout terminal.

Each of the store's employees has a PDA that's linked, via a wireless local area network (WLAN), to all other on-site PDAs and to back-end data. Other featured technologies include electronic advertising displays, shopping cart-mounted touch screens that direct customers to specific products, multimedia information kiosks, and electronic shelf labels that can be instantly updated. Project partners include IBM, Intel, Philips, SAP, and over 30 other technology companies.

The store's goal is to test new retailing technologies and to set standards that can be implemented on an international scale, says Albrecht von Truchsess, a METRO Group representative. "We want to practice how a variety of technological systems can work together in a very complex way." He notes that new technologies will be added to the store whenever the company deems they are ready for public testing.

METRO doesn't plan to open any additional Future Stores. "It's a test lab," notes von Truchsess. "You normally don't erect several test labs." But that doesn't mean that technologies tested in the Future Store won't eventually find their way into the retail mainstream. "We will get results from this store, and we will decide which solutions are fit to be brought into other stores," says von Truchsess.

6.5 RADIO MONITORS

Unlike RFID tags, which only provide tracking capabilities, inexpensive monitoring circuits allow enterprises to carefully supervise the condition of key assets. Stockholm, Sweden-based Cypak, for instance, has developed a sensor-based monitoring technology that's aimed at product delivery surveillance and control. Using special conductive inks, Cypak prints a microprocessor, environmental sensors, antenna, and support electronics directly onto product and shipping packages. The company's SecurePak technology stores a unique identification that can be programmed with unalterable information about the package's source, destination, and contents. The device can then record whether the package it's attached to has been opened, closed, or tampered with in any way during shipping; by communicating with external readers, it can even tell shippers when and where such incidents occurred. The circuitry adds about $2 to a package's cost, notes Jakob Ehrensvärd, Cypak's CEO. "This is basically a chip on a sticker," he says. The reader presently costs about $10.

The Swedish Postal Service recently tested SecurePak for shipping high-value products, such as computer equipment, precious artworks, and government documents. Thord Axelsson, the agency's chief security officer, says the technology allows postal employees to almost instantly determine when and where a package has been opened, rather than waiting the one to two weeks that a manual investigation would require.

SecurePak can even tell its user exactly how a package was tampered with, for example, if it was opened indirectly or if a knife was used on the package. The device's sensors also allow users to detect whether shipments have been exposed to potentially damaging environments, such as extreme heat or cold or traumatic shocks. Axelsson says the technology is cheap enough to be disposable, yet rugged enough to be reused several times. Although Axelsson was initially dubious that such a small, inexpensive technology could provide so much information, the recent tests proved SecurePak's worth. "We can see now that they are working," he says.

Cypak's disposable "cardboard computer" technology can also be used in pharmaceutical packaging. When a patient breaks open a blister pack to take a pill, the monitoring circuit records the date and time. The data is then read from the packaging when the user visits his or her doctor. "It confirms whether the patient is following the doctor's instructions," says Ehrensvärd.

6.6 VEHICULAR TELEMATICS

Telematics—vehicle-based information systems—will make great advances over the next couple of decades. Current telematics systems, such as General Motor's OnStar, already provide several basic support services, such as location-based news and weather reports and the ability to remotely unlock inadvertently locked doors. But Ken Hopkins, director of product innovation for Farmington, Michigan-based Motorola Automotive, believes that that many more exciting telematics services are just down the road. He predicts that by 2025 vehicles will be able to drive themselves without human interaction, providing a much safer traveling environment as well as a true mobile office. "In other words, you'll get in, tell the vehicle a destination, and it will get you there," he says. In the meantime, Motorola is developing collision-avoidance technology that will automatically alert drivers to upcoming obstructions, such as debris in the road or another vehicle ahead. "We're going to enter into an era where we can actually prevent accidents," he says.

The widespread acceptance of global position system (GPS) technology will be critical to telematics' overall growth. As consumer awareness of GPS increases, so will product innovation and total market revenue. Roughly half of the GPS market today consists of automotive and asset-tracking equipment. These segments will still continue to grow at rates faster than that of the broader market for GPS equipment, according to ABI, a technology market research firm located in Oyster Bay, New York.

Despite the strength of these markets, new GPS applications are constantly emerging, for example, people-tracking devices and GPS golf systems. The net result will be a GPS market worth over $22 billion by 2008, according to ABI figures. Companies like Garmin, Wherify Wireless, and Navman are synonymous with integrating GPS receivers into innovative form factors. Advances in GPS integrated circuits (ICs) will fuel this trend across the entire industry. Sony's recent unveiling of a miniature, single-chip IC provides further evidence that more of these novel applications are likely in an ever-increasing range of devices.

6.6.1 Vehicular Radar

Futurists have long predicted the creation of a vehicle-based radar system that would allow cars, trucks, and buses to safely avoid obstacles, even in zero-visibility conditions. High cost and bulky equipment have so far frustrated numerous attempts to create a practical vehicle radar, but help may soon be on the way. That's because California Institute of Technology researchers have built the world's first radar on a chip, implementing a novel antenna array system on a single, silicon device.

The chip contains both a transmitter and receiver (more accurately, a phased-array transceiver) and works much like a conventional antenna array. But unlike conventional radar, which involves the mechanical movement of hardware, Caltech's as-yet-unnamed chip uses an electrical beam that can steer the signal in a given direction in space without any mechanical movement. In cars, an array of the chips—one each in the front, back, and right and left sides—could provide a smart cruise control. Such a system wouldn't just keep the pedal to the metal, but would brake for a slowing vehicle ahead, avoid a car that suddenly dodges in front, or evade an obstacle that suddenly appears in the vehicle's path.

There are other radar systems in development for cars, but these consist of a large number of modules that use more exotic and expensive technologies than silicon. The Caltech chip could prove superior because of its fully integrated nature, which allows it to be manufactured at a substantially lower price and makes the chip more robust in response to design variations and changes in the environment, such as heat and cold. "Traditional radar costs a couple of million dollars," says Ali Hajimiri, an associate professor of electrical engineering at Caltech and the project's leader. "It's big and bulky and has thousands of components. This integration in silicon allows us to make it smaller, cheaper, and much more widespread."

Silicon is the ubiquitous element used in numerous electronic devices, including the microprocessor inside personal computers. It is the second most abundant element in the earth's crust (after oxygen), and components made of silicon are cheap to make and are widely manufactured. "In large volumes, it will only cost a few dollars to manufacture each of these radar chips," says Hajimiri. "The key is that we can integrate the whole system into one chip that can contain the entire high-frequency analog and high-speed signal processing at a low cost," he notes. "It's less powerful than the conventional radar used for aviation, but since we've put it on a single, inexpensive chip, we can have a large number of them, so they can be ubiquitous."

The chip also has several other telecom applications. For communications systems, the chip's ability to steer a beam allows it to provide a clear signal. Mobile phones, for example, radiate their signal omnidirectionally. That's what contributes to interference and clutter in the airwaves. "But with this technology you can focus the beams in the desired direction instead of radiating power all over the place and creating additional interference," says Hajimiri. "At the same time you're maintaining a much higher speed and quality of service."

The chip can also serve as a wireless, high-frequency communications link, providing a low-cost replacement for the optical fibers that are currently used for ultrafast communications. The chip runs at 24 GHz (24 billion cycles in one second), an extremely high speed, which makes it possible to transfer data wirelessly at speeds available only to the backbone of the Internet (the main network of connections that carry most of the traffic on the Internet). A small device based on the Caltech chip could, for example, be placed on the roof of a house or office building, replacing bulky satellite dishes or cable/DSL connections.

6.6.2 Train Monitor

Cars, trucks, and buses aren't the only vehicles that stand to benefit from telematics technology. A new telecom-equipped monitor could lead to safer train travel.

Developed by a U.K. father and son team working at the University of Newcastle-upon-Tyne's Stephenson Center, Microlog is an advanced miniature data logger. The device, which is installed on a train's wheel axles, is able to detect any abnormal stresses that could be caused by problems on the track—buckling due to excessive heat, for example. Information collected by the unit can also help engineers better understand wheel-to-rail interaction and establish more reliable codes for future axle designs.

Microlog gathers the relevant data via sensors and uses satellite technology to detect the exact location of problem spots. It then uses GSM phone technology to send a warning message to a computer miles away. The remote software analyzes the data and alerts the train's operator to any problem that requires urgent troubleshooting. Microlog can also be remotely accessed and reprogrammed using a short-range radio link, the Internet or via the GSM network. The monitor packs 4MB of memory, a 16-bit microprocessor, and GPS and GSM technology into a case only one-third of the size of a matchbox.

"Although data loggers have been used for more than two decades, they have always been relatively big and their use has therefore been limited," says lead researcher Jarek Rosinski, who developed Microlog with his 18-year-old son, Martin, a University of Newcastle student. "Microlog is unique because of its miniature size, which allows us greater flexibility and means we can fit it to smaller components such as train axles," says Rosinski. "We have been working over several years to perfect the design and we are confident it has huge potential in a variety of applications, rail safety being just one of them." Other Microlog applications include troubleshooting power plants, automotive and marine transportation monitoring, and research and development data gathering.

Microlog is the product of several years of development by scientists associated with the University of Newcastle's design unit, one of six outreach business consultancies that are known collectively as the Stephenson Group. The group takes its name from Robert Stephenson, the 19th century entrepreneur who built the groundbreaking Rocket locomotive in a nearby Newcastle factory with his father, George. Testing on Microlog will start on the GNER East Coast Main Line, a route that Stephenson was involved in developing almost two centuries ago.

6.6.3 Satellite Road Tolls

Satellite technology already helps motorists find locations and plan routes. Within a few years, GPS and similar systems could also be used to automate the collection of road tolls and insurance payments for European drivers.

In an effort to make toll collection and car insurance rates more "fair," the European Space Agency (ESA) is looking into a technology that would implement "satellite-assisted distance pricing." The ESA has tapped Mapflow, an Irish provider of location technology products, to undertake a feasibility study to examine the possibility of implementing a Europe-wide road tolling system.

The research aims to establish whether satellite technology can be used to calculate the cost of motoring.

A real-world demonstration of the virtual tolling concept is planned to take place in Lisbon beginning in late 2004. Also under ESA funding, the Lisbon project is being conducted by the Portuguese company Skysoft in close cooperation with the Portuguese motorway authority.

Last April, the European Commission published a proposal that all vehicles should pay road tolls electronically, with full implementation foreseen for 2010. Under the proposal, all vehicles would carry a "black box" and would be tracked by satellites relaying information on the distance traveled by the vehicle, the class of road traveled, and the time at which the journey was made.

The research, commissioned by ESA on behalf of the European Union (EU), aims to evaluate the feasibility of a standard tolling approach throughout Europe. The study will look at the effects of such a system on Europe's road infrastructure as well as associated technology impacts.

Potential benefits of a harmonized road tolling system, according to the ESA, would be fairer toll and insurance fee implementation by charging on a "pay for use" basis, lower road building and maintenance costs as the need for physical infrastructure is reduced, and also lessened road congestion. Germany recently received EU approval to implement a new tolling system for commercial vehicles. The system currently being tested uses the U.S.-operated GPS technology. The German government hopes to raise 650 million euros a year through the new charges.

EU-wide satellite-assisted tolling would make use of Galileo, Europe's planned satellite navigation system. Galileo is a joint initiative between the European Commission and ESA to develop a global navigation system, scheduled to be operational by 2008.

The system will have a constellation of 30 satellites revolving in three circular medium earth orbits, approximately 24,000 km above the earth. This will create a network covering the entire globe, relayed at ground level by stations monitoring the satellites and the quality of their signals.

Once operational, Galileo will provide a highly accurate, guaranteed global positioning service under civilian control. It will be interoperable with other global satellite navigation systems, such as GPS, while providing greater accuracy, down to two meters. Other applications for Galileo in the transport sector include vehicle location, taxi and truck fleet management, and monitoring levels of road use.

6.7   HELPING RANCHERS FROM SPACE

Satellite technology can also be used to track things other than vehicles, including ocean currents, migratory animals, and crops. In fact, powerful new satellite imaging database software is coming to the aid of North American ranchers and other people who work on the land.

University of Arizona researchers have created a Web database that allows ranchers to compare land greenness from one year to the next, between years, against a 14-year average, and at two-week intervals. Such information can be invaluable for making long-term land management decisions.

Ranchers, forest rangers, and other natural resource managers work directly on the land nearly every day to observe changes and decide how to handle them, whether grazing cattle, monitoring wildlife, or assessing fire danger. A new University of Arizona satellite image database, called RangeView, offers users a bird's-eye view of broad terrain. "RangeView provides frequent satellite images online to enhance the ability of natural resource managers, including ranchers, to manage the landscape," says Chuck Hutchinson, director of the Arizona Remote Sensing Center, part of the Office of Arid Lands Studies at the University of Arizona's College of Agriculture and Life Sciences.

Stuart Marsh, professor of arid lands, and arids-lands researcher Barron Orr, and Hutchinson created the Web site to display NASA digital images in configurations that allow users to analyze the characteristics of the land. "This tool offers the ability to zoom in on your ranch, forest, or habitat and monitor changes in vegetation through time," says Orr. The RangeView Web site provides applications for viewing, animating, and analyzing satellite images to monitor vegetation dynamics through time and across landscapes. The site also offers a step by step tutorial for new users.

The database can be used to find answers to a variety of questions. Hitting the "animate" button to see two-week variations in vegetation over the past year, for example, enables users to assess fire potential or other time-dependent applications. To orient themselves, viewers can look at the location of towns, roads, grazing allotments, and other features. "Our members are so familiar with the land. When you get them looking at a view of it from space, they forget they're using a computer mouse and want to get that cursor arrow on their ranch," says Doc Lane, director of natural resources for the Arizona Cattlemen's Association and Arizona Wool Producers Association. "You'd be amazed how quickly people forget they're using a computer and connect directly with their ranch on the screen."

Resolution on the satellite images is one square kilometer, enough to show vegetation color without violating people's privacy. Hutchinson says the capability of the site is somewhere between weather information that is not very site-specific and field monitoring that is quite site-specific. "This is something in the middle that can bridge those two scales." RangeView images are available for the entire United States, southern Canada, and northern Mexico.

6.8 SEEING INSIDE WALLS

The building community soon may have radio vision—a new way to "see" moisture inside walls. Building researchers at the National Institute of Standards and Technology (NIST) have joined forces with Intelligent Automation

Inc. in Rockville, Maryland, to develop a way to use ultra-wide-band radio waves to nondestructively detect moisture within the walls of a building. As any homeowner who's suffered with leaky plumbing or mold problems will tell you, the present state of the art for pinpointing moisture problem areas relies mostly on guesswork and a drywall saw.

Based on hardware developed by Intelligent Automation, the new NIST technique involves sending a broad range of radio frequencies through typical drywall construction to look for a "moisture" signature in the signal that is reflected back. Laboratory experiments conducted with a simplified wall section made of gypsum board, fiberglass insulation, and oriented strand board (similar to plywood) demonstrated that the new method can locate moisture pockets to within one centimeter.

The presence of water within the model wall produced a stronger reflection of radio waves at specific frequencies. The elapsed time between transmission of the waves and their arrival at a receiving antenna helps determine the location of the water. By processing the reflected signals with computer software, the researchers can create detailed three-dimensional maps that highlight wet areas.

Research is continuing to see how well the apparatus performs with real walls that include studs, wires, pipes, and windows, which may complicate the readings.

6.9 MICROSCILLATOR

Mobile phones and other radio-based devices could be created with the help of a new microscillator. The tiny, novel device for generating tunable microwave signals has been developed by researchers at NIST. The device measures just a few micrometers square and is hundreds of times smaller than typical microwave signal generators in use today in cell phones, wireless Internet devices, radar systems, and other applications.

The technology works by exploiting the fact that individual electrons in an electric current behave like minuscule magnets, each one with a "spin" that is either up or down, just as an ordinary magnet has a north and a south pole. The device consists of two magnetic films separated by a nonmagnetic layer of copper. As an electric current passes through the first magnetic film, the electrons in the current align their spins to match the magnetic orientation in the film. But when the now aligned electrons flow through the second magnetic film, the process is reversed. This time the alignment of the electrons is transferred to the film. The result is that the magnetization of the film rapidly switches direction, or oscillates, generating a microwave signal. The microwave signal can be tuned from less than 5 GHz (5 billion oscillations a second) to greater than 40 GHz.

The NIST experiments confirm predictions made by theorists at IBM Corp. and Carnegie Mellon University in 1996. NIST physicist William Rippard says the new oscillators can be built into integrated circuits with the same technologies now used to make computer chips and that they may eventually replace bulkier technologies at a greatly reduced cost.

6.10 ANTENNA TECHNOLOGIES

For most people, even many engineers, the antenna is the least interesting part of a radio. Yet radio performance depends greatly on the type of antenna used. In fact, the best transmitter or receiver is little more than a high-tech doorstop unless it's connected to a high-quality antenna. That's why antenna research is critical to the future of radio technology.

6.10.1 High Dielectric Antenna

Researchers are working on several new antenna technologies. One promising design, which could significantly boost 3G wireless data transfer speeds, is being developed by Cambridge, United Kingdom-based Antenova. The device breaks all the normal rules of antenna design, yet improves signal quality so much that handsets can actually support services such as playing real-time video clips—a feat that has proved elusive to date.

The breakthrough comes from Antenova's high dielectric antenna (HDA) technology, which enables two or more antennas to listen independently even though they are mounted within millimeters of each other. The antenna also breaks new ground in size and efficiency. The antenna's reception sensitivity, which is based on an advanced ceramic material, is significantly improved compared with a conventional copper antenna, and antenna dimensions are so small that they're easily mounted inside a handset.

"The antenna is a cornerstone of all wireless products, yet the technology has changed little in decades and remains inefficient, inflexible, and problematic to design", says Colin Ribton, an Antenova spokesman. "High dielectric antennas completely transform the situation, opening up unprecedented new potential for all wireless products—from phones and terminals to LAN and base station infrastructure."

According to Antenova, HDA not only improves the size and performance of antennas, it enables completely new design possibilities. These include the first practical approach to incorporating "antenna diversity" into products, greatly enhancing signal strengths and with it data communications performance. Directionality is another possibility, which may be used to foster coexistence of multiple protocols on equipment, to add spatial multiplexing of spectrum, to increase the capacity of wireless networks, and to reduce electromagnetic radiation directed toward users. Antenna diversity would also allow broadband antennas to be specifically designed to mobile phones or in extremely narrowband versions to eliminate the conventional need for external pass-band filters. The antennas themselves can be as small as one-tenth the size of a conventional product, as well as much more sensitive. Efficiencies as high as 70 or 80 percent are easily achievable, claims Antenova.

Antenna diversity simply means using two or more antennas rather than one, says Ribton. "With this form of antenna, next-generation phones can actually achieve the headline-making 3G services everyone has been talking about for years," he says. "Our trials show that two antennas can improve data reception speeds by fourfold. That equates to the difference between making an ordinary voice call and experiencing a video conference."

Antenova has also constructed a tool suite, including its own simulator, which allows high-performance antennas to be created rapidly to suit the specific needs of product developers. The company's own anechoic chambers and other test facilities allow samples to be provided rapidly to clients for evaluation and product prototyping. The design can be provided for OEMs to manufacture themselves or produced by Antenova in conjunction with its volume-manufacturing partners. Antenova says it is also working on HDA technology with developer partners engaged in almost every segment of the wireless industry.

6.10.2 Nanotube Antenna

Tiny antennas, thousands of times smaller than the width of a human hair, could soon lead to better performing mobile phones and other radio-based devices.

Research at the University of Southern California (USC) shows that minuscule antennas, in the form of carbon nanotube transistors, can dramatically enhance the processing of electrical signals. "No one knows exactly how these little tubes work or even if they will work out in manufacturing, but they are surprisingly good at detecting electrical signals," says Bart Kosko, a professor in USC's electrical engineering department and the project's lead researcher. "Once we figure out all the parameters that are needed to fine tune them, both physically and chemically, we hope to turn these tubes into powerful little antennas." If all goes well, the tubes could start appearing in consumer products within five to ten years, Kosko predicts.

Kosko's research is focused on a well-known but counterintuitive theory called "stochastic resonance" that claims noise, or unwanted signals, can actually improve the detection of faint electrical signals. Kosko set out to show that the theory was applicable at the nanoscale. Under controlled laboratory conditions, graduate student Ian Lee generated a sequence of faint electrical signals ranging from weak to strong. In combination with noise, the faint signals were then exposed to devices with and without carbon nanotubes. The signals were significantly enhanced in the container with the nanotubes compared with those without nanotubes, Kosko says.

Although additional testing must be conducted before the structures can be of practical use, Kosko sees a big potential for the little tubes. He says they show promise for improving "spread spectrum" technology, a signal-processing technique used in many newer phones that allows listeners to switch to different channels for clearer signals and to prevent others from eavesdropping. Arrays of the tiny tubes could also process image pixel data, leading to improved television images, including flat-panel displays, according to Kosko. The tubes also have the potential to speed up Internet connections.

In a more futuristic application, Kosko believes the tubes have the potential to act as artificial nerve cells, which could help enhance sensation and movement to damaged nerves and limbs. The sensors might even be used as electrical components in artificial limbs, he notes.

By adjusting the shape, length and, chemical composition of the nanotubes, as well as the size of the tube array, the devices can, in essence, be customized for a wide variety of electronic needs. "There are likely many good applications for the technology that we have not foreseen," says Kosko.

6.10.3 Fractal Antennas

Antennas for the next generation of mobile phones and other wireless communications devices may bear a striking resemblance to a mountain range or perhaps a rugged coastline or forest.

University of California Los Angeles (UCLA) researchers are using fractals—mathematical models of mountains, trees, and coastlines—to develop antennas that meet the challenging requirements presented by the more sophisticated technology in new mobile phones and mobile communications devices. Antennas for these systems must be miniature and they must be able to operate at different frequencies, simultaneously. "Manufacturers of wireless equipment, and particularly those in the automotive industry, are interested in developing a single, compact antenna that can perform all the functions necessary to operate AM and FM radios, cellular communications, and navigation systems," says Yahya Rahmat-Samii, who chairs the electrical engineering department at UCLA's Henry Samueli School of Engineering and Applied Science.

Fractals, short for "fractional dimension," are mathematical models originally used to measure jagged contours such as coastlines. Like a mountain range whose profile appears equally craggy when observed from both far and near, fractals are used to define curves and surfaces, independent of their scale. Any portion of the curve, when enlarged, appears identical to the whole curve—a property known as "self-symmetry." Rahmat-Samii found that the mathematical principles behind the repetition of these geometrical structures with similar shapes could be applied to a methodology for developing antenna designs. Using this method, he has developed antennas that meet two important challenges presented by the new generation of wireless devices. They conserve space and can operate simultaneously at several different frequencies.

Fractal methodology allows Rahmat-Samii to pack more electrical length into smaller spaces, he says. Increased electrical length means that the antennas can resonate at lower frequencies. Because fractal designs are self-symmetrical (repeat themselves), they are effective in developing antennas that operate at several different frequencies. "One portion of the antenna can resonate at one frequency while another portion resonates at another frequency," notes Rahmat-Samii.

6.10.4 Fractal Antenna Design

At Pennsylvania State University, researchers are beginning to use fractal theory to design real-world antennas. Engineers have discovered that an antenna design, composed of an array of fractal-shaped tiles, promises to provide a 4:1 to 8:1 improvement in bandwidth compared with their conventional counterparts. This concept could lead to a new generation of base station antennas and other radio antennas that are smaller and can operate across a wider range of frequencies.

Many natural objects, such as tree branches and rivers and their tributaries, are versions of mathematical fractals. These designs appear pleasingly irregular to the eye but actually comprise self-similar, repeated units. The new broadband antenna design, developed by the Penn State engineers, is composed of irregular but self-similar, repeated fractal-shaped unit tiles, or "fractiles," that cover an entire plane without any gaps or overlaps. The outer boundary contour of an array built of fractiles follows a fractal distribution.

Compared with conventional antennas, fractal antennas have a smaller cross-sectional area, offer wider bandwidth, work without an impedance matching network, and provide higher gain. Building on the fractal antenna concept, Douglas H. Werner, a professor of electrical engineering and senior scientist in Penn State's Applied Research Laboratory, Waroth Kuhirun, Penn State graduate student, and Pingjuan Werner, a Penn State associate professor of electrical engineering have developed a design approach for broadband phased array antenna systems that combine fractal geometry with aspects of tiling theory.

The Penn State design exploits the fact that fractal arrays are generated recursively starting from a simple initial unit. Using this characteristic, the researchers were able to develop fast recursive algorithms for calculating radiation patterns. Leveraging recursive growth, the researchers were also able to develop rapid algorithms for adaptive beam forming, especially for arrays with multiple stages of growth that contain a relatively large number of elements. "The availability of fast beam-forming algorithms is especially advantageous for designing smart antenna systems," notes Werner.

Ultimately, the researchers discovered that a fractile array made of unit tiles based on the Peano-Gosper curve, a specific type of fractal, offers a performance advantage over a similarly sized array featuring conventional square boundaries. A Peano-Gosper fractile array produces no grating lobes over a much wider frequency band than conventional periodic planar square arrays. "Grating lobes are sidelobes with the same intensity as the mainbeam," explains Werner. "They are undesirable because they take energy away from the main beam and focus it in unintended directions, causing a reduction in the gain of an antenna array."

Penn State is in the process of patenting the team's approach to Peano-Gosper and related fractile arrays. The team has also been awarded a grant through the Applied Research Laboratory to build and test a prototype antenna.

6.10.5 Towers in the Sky

Although antenna design and construction is important, so is placement. In radio, generally speaking, the higher an antenna is placed, the better it will function. That's why satellites work so well. But satellites are very expensive to build and launch. That's why a couple of companies are aiming their sites somewhat lower.

"A big problem the U.S. has is that 80 percent of the population lives in just 10 percent of the land," says Knoblach, citing 2000 U.S. Census figures. "If you build only towers, you could never build the number needed to cover that sparsest 20 percent of the country because of economics."

Space Data's SkySites project plans to launch radio transceivers connected to latex, expendable weather balloons from 70 sites in 48 states by the end of 2004. The balloons, which are 6 to 8 feet in diameter at launch but grow to about 25 to 30 feet in the air, are designed to float to 20 miles above the earth's surface, and will provide communications services for approximately 12 to 24 hours. At that point, they'll be replenished with a new "constellation" of balloons.

The company claims it can launch its weather balloons twice a day, year-round for about $15 million to $20 million a year, significantly less than what most carriers would spend on towers to fill the same coverage area. Space Data's system works with an array of existing wireless devices, including cell phones, pagers, and handheld computers. Telephone carriers will be able to offer service to rural customers with a cost structure previously not possible with conventional tower-based infrastructure.

The Space Data network will also help wireless service providers improve customer service by increasing their coverage. "By filling in all the coverage gaps, we can provide service everywhere," says Knoblach. The company plans to bring cell phone service to remote locations—such as Indian tribal lands—for approximately the same price customers currently pay for cell phone service in places like Phoenix and Tucson. "The cost we're targeting is really the same kind of cost that you pay per month now," explains Knoblach. Although initially costs would be a bit more expensive, as they ramp up, Space Data expects costs to drop closer to rates paid by urban customers.

Knoblach believes that his company is entering a wide-open market. Space Data's primary competitors—cell phone tower operators—face the huge financial burden of building sites in isolated locations that may receive only a few phone calls per month. The company's only other competitors, satellite service providers, have their own problems. "They've largely gone bankrupt," notes Knoblach. "Although their satellites are installed, they're aging and getting close to the end of life. They haven't got a big enough business to actually fund the new system to replace them."

The idea of developing a balloon that functions as a communications satellite captured Knoblach's imagination in the mid-1990s. In 1996, Knoblach joined together with Eric Frische, a scientist he met while studying as an undergraduate at the Massachusetts Institute of Technology. The duo began the venture by laying down the project's technical groundwork. By May 2000, Knoblach and Frische were ready to hire a staff and begin flight testing. "We've been in the R&D phase from then until now," Knoblach says. Frische would eventually become Space Data's vice president of technology and chief technology officer.

Outside of technical issues, a big part of Space Data's work over the past several years has been ironing out federal regulatory issues with the Federal Aviation Administration and the Federal Communications Commission (FCC). "We now have a waiver from the FCC to allow us to operate this system like commercial wireless spectrum," says Knoblach. Space Data has also acquired 1.4 MHz of frequency space, at a cost of $4.2 million, in the 900-MHz band, necessary for transmitting cell phone signals from the earth to the balloons and back again. Eventually, Space Data wants to deploy SkySites internationally, supporting both voice and data communications.

Another company looking toward balloon-based communications is Global Aerospace Corp. The company is planning a new type of satellite that could provide communications in remote areas of the world with no technological infrastructure. The technology, which it has dubbed "Stratospheric Satellites," consists of NASA-developed "super-pressure balloons" that fly at 110,000 feet (Fig. 6-1). Combined with steering systems and a solar array used for power, the balloons can carry payloads up to the size and weight of a small truck.

Global Aerospace has developed a trajectory control and solar array system for the superpressure balloons, allowing them to be steered over disaster areas and powered over the course of its long life. At a current cost of $1.75 million per unit for development, and a projected life-cycle cost of $500,000 or less per unit in production, the Altadena, California-based company believes that Stratospheric Satellites can be a low-cost alternative to remote communications platforms provided by aircraft and space-based satellites. Because Stratospheric Satellites fly much closer to Earth than space-based satellites, they provide 20 times higher resolution surface images of disasters and 160,000 times higher signal radar than space-based satellites.

"Because they are relatively inexpensive, can be steered, are independently powered, and can carry a large payload, they will probably be the most cost-effective way of bridging the last mile in telecommunications coverage," says Kerry Nock, Global Aerospace's president. Nock estimates that a constellation of 400 Stratospheric Satellites covering most of the populated areas in the

Figure 6-1 *Stratospheric satellites.*

northern hemisphere would cost less than $100 million—less than the cost of a single space-based satellite including its launch. Nock belies that operational costs will come in at less than $10 million per year.

6.11 INTERFERENCE

It's impossible to have a chapter on radio technologies without including some information on interference. Whether natural or man-made, interference has been an integral—although unfortunate—part of radio since the days of crystal sets. At the dawn of the 21st century, researchers are working hard to remedy one of radio's oldest problems.

6.11.1 An Allocation Approach

Penn State University engineers have developed an economical way to more efficiently manage radio spectrum use and prevent interference on wireless broadband systems for high-speed Internet access. "With this technique, service providers could offer quality service to more homes using only a limited span of the radio spectrum," says Mohsen Kavehrad, director of Penn

State's Center for Information and Communications Technology Research. "If providers can squeeze more customers onto the available bandwidth, it could translate into lower costs for the consumer." The approach also promises equipment cost savings since simulations show that the new scheme maintains performance at top industry standards with more economical components.

Currently, high-speed Internet access capable of carrying MP3 files, video, or teleconferencing is available primarily over wired networks. However, wireless local loops are being introduced as broadband alternatives in some test markets. These new wireless networks are facing serious obstacles in competing for bandwidth, sometimes having to share bands with cordless phones or even microwave ovens. Even when the wireless providers use licensed bands, they face the prospect of many customers simultaneously uplinking and downlinking information across the net, creating cochannel interference.

The wireless local loops work much like cell phones via a base station that sends the radio signals carrying the Internet connection out to any customer whose residence or business is equipped with an appropriate antenna. Unlike cell phone usage, however, the two-directional uplink and downlink traffic between the customer and the Internet provider's base station is more asymmetrical with very little use during sleeping hours and lots of use when kids come home from school and download music or play games, for example. Kavehrad says, "The nature of multimedia traffic is not static in uplink and downlink directions, as with voice telephony, and the bandwidth is more biased toward downlink transmissions."

Wireless local loops need both software and hardware that enables the network to respond to the changes in traffic while also making sure that every hertz in the available spectrum is used as efficiently as possible. In addition, the system must contend with the fact that some incoming interfering signals are stronger than others.

The solution developed by the Penn State engineers is software that allows the subscriber signal, whose direction of arrival is subject to a lesser number of strong interferers, to be processed ahead of the ones experiencing the most interference. In other words, the new strategy is a scheme that allows avoiding strong cochannel interference by sequencing the processing of the signals according to the amount of interference they are experiencing. Because the amount of interference any subscriber's signal experiences varies microsecond by microsecond, no subscriber has to wait very long for a turn.

"The usual techniques employed to suppress interference use adaptive spatial filters, which require expensive RF components and a large number of computations to queue the subscribers' signals," says Kavehrad. "However, with our approach, we need only a simple, cost-effective spatial filter and relatively fewer computations. Our simulations show that the performance of the new approach and the traditional technique are comparable. Thus, our strategy shows a practical compromise between complexity and cost, while achieving the desired signal quality."

6.11.2 Quieter Ovens

As just about anyone who has ever used a wireless phone or WLAN can attest, wireless ovens are almost as good at creating electrical interference as they are at heating up TV dinners. Researchers at the University of Michigan College of Engineering pondered this problem and have developed an elegantly simple technique that dramatically reduces the interference microwave ovens create in telephones and wireless computer networks.

Worldwide, there are hundreds of millions of microwave ovens in kitchens, offices, and laboratories, each with a magnetron that creates communications problems ranging from an aggravating crackle during a friendly telephone call, to the disruption of 911 calls and the flow of data in wireless computer networks. Although these effects are difficult to quantify, it's safe to say they're an annoyance, an economic drain, and potentially life-threatening.

The basic difficulty is that microwave ovens operate at a frequency near 2.45 GHz—about the same frequency at which telephones and wireless computers operate. In the microwave oven, there are two magnets, one at each end of the magnetron. In an amazingly simple discovery, University of Michigan College of Engineering Professor Ronald Gilgenbach and a research team composed of Professor Y. Y. Lau and graduate student, Bogdan Neculaes, all from the department of Nuclear Engineering and Radiological Sciences, found that when they added four permanent magnets to the outside of one of the standard magnets, they could disrupt the magnetic field in such a way that it becomes benign to nearby electrical devices, yet doesn't significantly affect the performance of the microwave oven.

The discovery could also have an enormous impact on the signal-to-noise ratio in radar, say the researchers.

Chapter 7

The Unblinking Eye—
Security and Surveillance

Besides allowing people to easily communicate with each other and to obtain vital information effortlessly, telecom also enables businesses, corporations, and other organizations and individuals to know what other people are doing. Emerging telecom technologies, including camera phones, networked surveillance cameras, and wireless security sensors, allow the monitoring of streets and property to an extent unimaginable only a few years ago.

On the other hand, telecom itself depends on security technologies. Without systems in place to safeguard devices and networks, telecom systems would be highly vulnerable to intruders seeking financial gain seeking to create havoc. This means that security is undoubtedly a double-sided proposition, protecting both people and the systems they use.

7.1 TESTING SECURITY

In the aftermath of September 11, 2001, the federal government has increased its interest in telecom-driven security and surveillance technologies and cybersecurity in general. Spearheading the government's security initiative at the academic level is the National Science Foundation (NSF). The NSF recently made 23 awards for 11 projects that will develop networking testbeds for research into cybersecurity, next-generation wireless and optical networking, and leading-edge scientific applications. These testbeds will let researchers push new networking technologies to the breaking point and beyond, paving the way for a more reliable Internet of tomorrow.

"Fundamental networking research has an essential role in advancing the country's digital and physical infrastructure," says NSF's Mari Maeda, acting division director for Advanced Networking Infrastructure and Research. "These testbed projects demonstrate how NSF contributes both to cutting-edge research and the next-generation networks we will depend on in our daily lives." Through NSF support, the testbeds, once deployed, will be open to experiments by networking researchers from other institutions.

To enhance the country's cyber defenses, the University of California (UC), Berkeley, and the University of Southern California have undertaken a large-scale testbed for experimenting with methods to protect networks against computer worms, viruses, denials of service, and other cyber attacks. With this testbed, attacks can be unleashed without a threat to the security of operational networks. A companion project led by researchers at Pennsylvania State University, UC Davis, Purdue University, and the International Computer Science Institute will develop scenarios for testing and evaluating proposed defense systems. To support these efforts, NSF collaborated with the Department of Homeland Security, which is cofunding a portion of $10.8 million awarded to these two projects.

The explosion of "wi-fi" networking, from coffee-shop hotspots to scientific networks of remote wireless sensors, is changing the dynamics of what used to be a primarily wired Internet. NSF has made awards to deploy five testbeds where researchers can develop technologies for next-generation wireless networks.

The testbeds will provide opportunities to benchmark new wireless protocols, evaluate prototype hardware, and examine emerging issues such as interference and efficient spectrum usage. These five testbed efforts each target a different aspect of wireless networking and are led by scientists at UCLA, Rutgers University, MIT, the University of Kansas, and the Stevens Institute of Technology.

Although wi-fi networking has emerged as a key technology for getting online, the links for network backbones and leading-edge scientific applications require the blazing performance of wired, optical networks. The $6.7 million DRAGON testbed, deployed by the University of Maryland, University of Southern California (USC), and George Mason University, will advance networking architecture research and bring about the intelligent control of optical transport networks.

A $3.5 million optical testbed deployed by the University of Virginia, Oak Ridge National Laboratory, the City University of New York's City College, and North Carolina State University will allow development and testing of cutting-edge science applications.

For networking research that requires the most realistic networked environments, two additional projects totaling $8.8 million will support ongoing testbed activities. In the first such project, the University of Utah and the

University of Wisconsin, Madison, will link Utah's Netbed and the Wisconsin Advanced Internet Laboratory in a testbed that can emulate more than 150,000 wired and wireless connections between tiny wireless sensors, personal computers, routers, and high-end computing clusters.

The second ongoing testbed activity, PlanetLab, is being supported by NSF awards to Princeton University, the University of Washington, and UC Berkeley. PlanetLab is a worldwide network "overlay" that runs on top of, but doesn't disrupt, existing networks and lets networking researchers conduct experiments at the scale of, and with the unpredictable behavior of, the global Internet.

7.2 LOCATION-BASED PRIVACY SOFTWARE

Location-based services promise custom-tailored access to a rapidly increasing number of connected activities. Yet existing location-based services don't support the possibility that a user's willingness to share location data may depend on a range of factors, including recent and current activities, time of day, and who is requesting the data—in other words, an individual's context.

Responding to this shortcoming, researchers from Lucent Technologies' Bell Labs have developed new software technologies that promise to enable users to tightly control how their location information is shared when using location-enabled mobile devices, such as mobile phones and PDAs. The software, when taken as a whole, provides a framework that will allow users to specify what location information is shared as well as when, with whom, how, and under what circumstances.

Bell Labs' Privacy-Conscious Personalization (PCP) framework relies on user preferences to intelligently infer context, such as working or shopping, and then determines with whom location information should be shared. When a user's location or other information is requested, it is analogous to someone making a query of a database. In PCP's case, the request is checked against the user's preferences and filtered through a high-performance rules engine, known within Bell Labs as "Houdini," before any action is taken. Because location and other mobile services require near-real-time performance, this entire process can take a few milliseconds or less.

Bell Labs' technology is based on the belief that privacy isn't a "yes or no" question but rather is a spectrum of possible answers based on a wide range of factors and user situations. Today's location-based services have limited capabilities—mobile users can choose to show their location to everyone, only to an explicitly selected group of buddies, or only to authorized officials in emergency situations using the government-mandated Enhanced 911 (E911). Currently available location tracking services are essentially "one size fits all"—users do not have the ability to tailor this capability to meet their individual preferences.

The PCP framework will enable consumer and business users to specify which individuals, groups, or businesses can see where they are based on the users' preferences. For example, during working hours, field sales representatives could opt to have their presence visible to their boss, but, after 5:30 p.m., location sharing with the boss can be disabled. Additionally, a sales rep might permit important customers to see his or her presence at anytime, but only with an accuracy of up to 10 miles. This capability would be especially helpful when a sales rep is visiting a key customers' competitor, for example.

As for the problem of unsolicited pitches from retailers, whether in the form of coupon spamming or short message service (SMS) advertisements, the PCP framework enables users to specify which kinds of businesses are allowed to see their location in a particular context, if at all. For example, a user may be interested in receiving a coupon from their favorite coffee retailer only when shopping, or on weekends before 9:00 a.m., or when within one mile of a golfing buddy. If none of these conditions holds, that user and his or her location would not appear to that retailer.

"The Bell Labs PCP framework promises to give mobile users the benefits they want from sharing location information without having to buy into a wholesale surveillance mechanism," says Jason Catlett, president of Junkbusters, a privacy advocacy firm located in Green Brook, New Jersey. "The fine-grained options, allowing the user to consent to disclosure of location according to place, time, and person are important to avoid being monitored 24/7."

A key challenge in the personalization of telecommunications services is the realization that various kinds of factors and preferences are relevant to different applications (e.g., sharing location information versus call forwarding) and different classes of users (e.g., office workers versus students, basic users versus power users). Bell Labs' PCP framework will allow network operators to preconfigure and offer different preference "palettes" tailored to these different applications and classes of users. The palettes could, for example, have drop-down menus prepopulated with the kinds of rules most relevant for specific classes of users, so they can quickly set their own preferences on personal computers or mobile phones.

"Bell Labs' technology would give end users more explicit control over how their network data is interpreted and shared with different requestors of location information in near real time," says Rick Hull, director of network data and services research at Bell Labs. "By offering powerful personalization capabilities, network operators can roll out new revenue-generating location services that also cater to the varying privacy needs and desires of different classes of users."

Service providers have generally acknowledged that having a policy engine to enforce privacy of user location information is a must-have capability. At the writing of this book, commercial trials of the PCP framework are being discussed with several carriers, and the "Houdini" rules engine is already being used by Lucent in prototype demonstrations of context-aware location-based services on mobile devices, as well as in preferences-driven call forwarding and blocking using both circuit-based and session initiation protocol (SIP) phones.

7.3 SECURING PRIVACY

As more smart phone, PDA, and PC users connect to servers in order to participate in shopping, banking, investing, and other Internet activities, a growing amount of personal information is being sent into cyberspace. Furthermore, every day, businesses and government agencies accumulate increasing amounts of sensitive data. Despite advances in cryptology, security, database systems, and database mining, no comprehensive infrastructure exists for handling sensitive data over their lifetimes. Even more troubling, no widespread social agreement exists about the rights and responsibilities of data subjects, data owners, and data users in a networked world. Now, collaborators in a new NSF project aim to bring order to the chaotic world of personal data rights and responsibilities.

The privacy project's goal is to invent and build tools that will help organizations mine data while preserving privacy. "There is a tension between individuals' privacy rights and the need for, say, law enforcement to process sensitive information," says principal investigator Dan Boneh, an associate professor of computer science and electrical engineering at Stanford University. For example, a law enforcement agent might want to search several airline databases to find individuals who satisfy certain criteria. "How do we search these databases while preserving privacy of people who do not match the criteria?" asks Boneh, who notes that similar questions apply to health and financial databases.

Government and business both want more access to data, notes Joan Feigenbaum, a Yale computer science professor and one of the project's investigators. She notes that individuals want the advantages that can result from data collection and analysis but not the disadvantages. "Use of transaction data and surveillance data need to be consistent with basic U.S. constitutional structure and with basic social and business norms," she says.

The project will join technologists, lawyers, policy advocates, and domain experts to explore ways of meeting potentially conflicting goals—respecting individual rights and allowing organizations to collect and mine massive data sets. They will participate in biannual workshops and professional meetings, collaborate on publications, and jointly advise student and postdoctoral researchers.

The researchers hope, for example, to develop tools for managing sensitive data in peer-to-peer (P2P) networks. Such networks allow hundreds, or even millions, of users to share data, music, images, movies, and even academic papers without the use of a centralized Web server. But participants' computers also may hold private files that users may not want to share.

Additionally, the researchers will explore ways to enforce database policies. For example, privacy-preserving policies need to be better integrated into database management systems to ensure compliance with laws such as the Health Insurance Portability and Accountability Act (HIPAA), Feigenbaum says.

The participants also hope to create a new generation of technology that can thwart what Boneh calls "the fastest growing crime in the U.S. and in the world"—identity theft, a substantial amount of which happens online. "Spoofed" Web sites, for example, pretend to be something they're not to entice users to enter sensitive information, such as a credit card or Social Security number. The spoofer can then use the information to apply for credit cards in the victim's name or otherwise usurp the individual's digital persona.

The privacy project is being funded by the NSF's Information Technology Research program. Research partners, who will receive $12.5 million over five years, are Stanford, Yale University, the University of New Mexico, New York University, and the Stevens Institute of Technology. Nonfunded research affiliates include the U.S. Secret Service, the U.S. Census Bureau, the U.S. Department of Health and Human Services, Microsoft, IBM, Hewlett-Packard, Citigroup, the Center for Democracy and Technology, and the Electronic Privacy Information Center.

7.4 THE SEEING EYE

The arrival of camera phones has led to widespread concern that people may be using the gadgets for more than just taking snapshots of their friends. Although evidence remains sketchy and largely anecdotal, it's likely that many camera phone owners are using the devices to steal business and personal secrets and to invade the privacy of innocent people. For example:

- The U.S. Air Force recently banned camera phones in restricted areas after the National Security Agency warned they could pose a threat to homeland security.
- Students are suspected of using the cameras to cheat on tests, bringing images of notes into the classroom.
- A 20-year-old Washington state man was charged with voyeurism after he slipped a cell phone camera underneath a woman's skirt as she shopped for groceries with her son.
- A strip club owner in Kansas City, Mo., came out swinging against the technology, threatening to smash photo cams with a sledgehammer to protect the privacy of his patrons and dancers.
- In Japan, where nearly half of all cell phones are photo phones, magazine publishers have become concerned about consumers who snap shots of pages that they like instead of buying the magazine.

7.4.1   Observation Camera

On the other hand, camera phones have legitimate applications, too. And whereas camera-equipped mobile phones are typically used to snap and send pictures of friends, the Nokia Observation Camera is designed to take pictures of people who may be anything but friendly.

As its name indicates, the Nokia Observation Camera is designed to operate as a security device—inside stores, warehouses, homes, and other places people might want to keep under constant surveillance. The unit can be activated in several different ways: by a programmed timer, input from a built-in motion sensor, or on the command of an incoming text message. To transmit images, the camera connects to an MMS-enabled phone.

The $400 camera sends JPEGs at a resolution of 640 × 480, 320 × 240, or 160 × 120. The camera operates on 900-MHz and 1.800-MHz wireless GSM networks. The camera, minus the phone, measures about 4.75 inches deep, 3.5 inches wide, and 1.75 inches high. The unit sits on a wall-mountable adjustable stand. The camera has its own SIM card and, therefore, its own phone number.

Although its unlikely that devices like Nokia's camera will ever become as popular as camera-equipped mobile phones, the technology is certain to appeal to people who want to keep an unblinking eye on their property. The camera can also be used to snoop on employees and household workers, such as babysitters. This can lead users into murky legal territory, however. On its Web site, Nokia warns, "Some jurisdictions have laws and regulations about the use of devices that record images and conversations in public or private areas and about the processing and further use of such data."

7.4.2   Surveillance Legality

Wayne Crews, director of technology studies at the Cato Institute, a public policy research organization located in Washington, isn't overly concerned by business and homeowners using surveillance cameras to protect their property. "A burglar doesn't have an expectation of privacy," he notes. "You're free to look out your front window at your neighbor's front lawn, but you can't go into his yard and look into his windows." Local jurisdictions, however, can limit the surveillance of people working inside a business or home, particularly if the workers are unaware that such monitoring is taking place.

Crews is also very worried about the increasing adoption of wireless camera technology by police departments and other government agencies, particularly when such cameras are connected into information databases. "The rules of the game are not written regarding the use of these kinds of perpetual surveillance capabilities," he notes. "It's going to be the privacy fight of the future."

On the other hand, Crews is supportive of individuals using phone cameras to record instances of unlawful government and corporate activities. "If the eyes are being turned on us by these kinds of technologies, we're turning the eyes right back on corporations and the government," he says.

7.4.3   Security Video Network

Wireless video sensor networks have the potential to significantly enhance national security and emergency response efforts. Assistant Professor Thomas Hou and Professor Scott Midkiff of Virginia Tech's electrical and computer engineering department are studying factors that affect network lifetime.

Composed of interconnected, miniature video cameras and low-power wireless transceivers that process, send, and receive data, wireless video sensor networks can provide real-time visual data in situations where accurate surveillance is critical, says Hou, the project's principal investigator. These networks can help reduce the impact of security breaches on the nation's infrastructure and improve the government's ability to prevent, detect, respond to, and recover from both manmade and natural catastrophes.

Hou and Midkiff are focusing on the issues of power use and network topology. Receiving, processing, and transmitting video information places a high demand on the batteries that supply power to a wireless video network. This poses a problem, particularly when networks are operated in remote locations. "A major challenge of our research will be maximizing the lifetime of networks using components with limited battery power," says Hou.

Hou and Midkiff believe that improving network topology—the arrangement by which network components are connected—is the key factor in maximizing power efficiency. "An analysis of power dissipation at video sensor nodes suggests that communication consumes significantly more energy than any other activity," Hou noted. "By adjusting the topology of the network, we can optimize the transmitter power of video sensor nodes and extend network lifetime."

The researchers will employ algorithms (mathematical problem-solving procedures) and techniques developed in the field of computational geometry to help determine the most beneficial topology adjustments. "Developing good solutions for these networking problems is the key to unlocking the full potential of a large-scale wireless video sensor network," Hou says. As part of the ITR project, Hou and Midkiff also plan to develop a software toolkit that will implement the topology control techniques that they discover.

7.4.4   Focusing on Precrime

A closed-circuit video camera designed to monitor a public place for criminal activity is hardly a new idea. But a video surveillance system that can forecast trouble in advance is something else indeed.

Researchers at London's Kingston University have developed a video surveillance system that has the ability predict a criminal incident well before it takes place. The technology, currently being tested in the London Underground's Liverpool Street Station, can pick up the first signs of a potential criminal event, such as a mugging or an imminent terrorist attack.

Running on a sophisticated image analysis program called Cromatica, the system detects unusual activity by recognizing preprogrammed behavioral patterns. The system is able to mathematically track a person's movements and then, if the individual starts acting suspiciously, signals a warning to a security service or police.

The unique technology monitors pedestrian flow and can highlight overcrowding. It can also be used to spot people selling tickets illegally or attempting to harm themselves. The project, being developed with a team of leading researchers from throughout Europe, is expected to help prevent many of the hundreds of injuries and incidents that take place on the London Underground every year.

Sergio Velastin, lead researcher at Kingston's Digital Imaging Research Centre, believes the system has an even wider potential for saving lives and cutting crime. He notes that it will eventually be capable of pinpointing an unattended object in a terminal, for example, highlighting who abandoned the package and where that person is in the building—meaning terrorists could be apprehended before leaving the complex. Velastin notes that the project has already attracted substantial funding from the European Union.

The technology marks an important break from conventional video cameras, which require a high level of concentration. Often nothing significant happens for long periods of time, making it difficult for the person keeping an eye on the camera to remain vigilant. "Our technology excels at carrying out the boring, repetitive tasks and highlighting potential situations that could otherwise go unnoticed," says Velastin.

Although Velastin believes the advances in identifying unusual behavior are a crucial step forward, he stresses humans are still essential when it comes to making the system work. "The idea is that the computer detects a potential event and shows it to the operator who then decides what to do, so we are still a long way from machines replacing humans," he says.

7.4.5 Smart Surveillance Camera Software

As we walk down streets, across parking lots, and through airports, cameras are watching us. But who's watching the cameras? In many instances, nobody. The cameras often simply serve as tools to record a scene. Nobody looks at the video unless a crime or other important event occurs.

This situation may soon change. Computer science researchers at the University of Rochester are looking to make surveillance cameras more useful by giving them a rudimentary brain. "Compared to paying a human, computer time is cheap and getting cheaper," says Randal Nelson, an associate professor of computer science and the software's creator. "If we can get intelligent machines to stand in for people in observation tasks, we can achieve knowledge about our environment that would otherwise be unaffordable."

Far from being an electronic "Big Brother," Nelson's software would only focus on things it was trained to look for: like a gun in an airport or the absence of a piece of equipment in a lab. Nelson has even created a prototype system that helps people find things around the house, such as where the keys were left.

In developing the technology, Nelson set about experimenting with ways of differentiating various objects in a simple black-and-white video image, the kind created by most surveillance cameras. The software first looks for changes within the image, such as someone placing a soda can on a desk. The change is immediately highlighted as the software begins trying to figure out if the change is a new object in the scene or the absence of an object that was there earlier. Over time, other methods have been developed, such as matching up background lines that were broken when an object was set in front of them. A later version of the software, which works with color cameras, takes an inventory of an object's colors, allowing an operator to ask the software to "zoom in on that red thing," for example. The software will comply, even though the soda can in the example is both red and silver and overlaid with shadows.

Nelson is also working on ways to get a computer to recognize an object on sight. One of the tasks he recently gave his students was to set up a game where teams tried to "steal" objects from each other's table while the tables were being monitored by smart cameras. The students would find new ways to defeat the software, and Nelson would then develop new upgrades to the system so it couldn't be fooled again.

Although a six-month-old baby can distinguish various objects from different angles, getting a computer to perform this task requires a formidable amount of processing, particularly if the object is located in a complicated natural setting, like a room bustling with activity. Unlike a baby, the software needs to be told a lot about an object before it's able to discern it. Depending on how complex an object is, the software may need anywhere from 1 to 100 photos of the object from different angles. Something very simple, like a piece of paper, can be "grasped" by the program with a single picture; a soda can may require a half-dozen images. A complex object, like an ornate lamp, may need dozens of photographs taken from different angles to capture all its facets. Nelson's software is able to handle this work within seconds. The software quickly matches any new object it sees with its database of images to determine what the new object is.

The smart camera technology has been licensed to PL E-Communications, a Rochester, N.Y.-based company that plans to develop the technology to control video cameras for security applications. CEO Paul Simpson is already looking into using linked cameras, covering a wide area, to exchange information about certain objects, be they suspicious packages in an airport or a suspicious truck driving through a city under military control. Even unmanned aerial reconnaissance drones can use the technology to keep an eye on an area for days at a time, noting when and where objects move. "We're hoping to make this technology do things that were long thought impossible—making things more secure without the need to have a human operator on hand every second," says Simpson.

7.4.6 Motion-Tracking Cameras

More homes and small businesses may soon be able to install affordable telecom-based surveillance cameras, thanks to a University of Rhode Island (URI) researcher's invention. The new device, created by Ying Sun, a URI professor of electrical engineering professor, allows a single inexpensive camera to automatically track moving objects in real time, eliminating the need to link together several cameras in order to thoroughly cover a specific area.

Using low-cost, commercially available hardware, the Automatic Image Motion Seeking (AIMS) system follows a moving object and keeps the target at the center of the field of view. "This [device] has broad impact for security surveillance because it eliminates the need to have a full-time guard watching a video screen," says Sun, who began developing the unit in 2002. "It's one intelligence level above any other existing system, and we've found the right compromise between speed and accuracy."

The unit is also inexpensive. Sun says the device can operate on a $30 webcam as well as on more sophisticated equipment. The device simply requires a motor-driven, pan-tilt camera mount and a processor. With low-cost equipment, the system could cost less than $300, including camera, making it ideal for use in homes and small businesses. Because it can track movements, one AIMS camera can be just as effective as several stationary cameras.

At a rate of 15 frames per second, the camera analyzes images for any motion. Once a moving object is found, it feeds that information to the camera mount to begin tracking the object as it moves. "We're working on adding 'behavior modifiers' to the system as well, so that once the camera identifies motion it can be programmed to continue to track a given size, shape, or color regardless of any other motion," Sun says.

Sun believes that a camera that can quickly track motion has a psychological effect on criminals. "If they see that the camera is following their movements, they may think that a security guard is manually operating the camera and is aware of their presence. It's likely that the criminal would then decide to go elsewhere." Besides property surveillance at places such as ATM machines, offices, warehouses, factories, and homes, the camera has applications for homeland defense, military uses, child monitoring, playground surveillance, border patrol, and videoconferencing. "Existing videoconferencing equipment requires the speaker to remain in one place in front of a stationary camera. With the AIMS camera people can walk around and the camera will automatically follow them," Sun says.

Sun's technology is based on an image-processing algorithm for real-time tracking. Because of the effectiveness and computational efficiency of the algorithm, the feedback control loop can quickly achieve reliable tracking performance. The algorithm is implemented in the Visual C++ language for use on a Windows-based PC, but the algorithm can also be configured to operate on an embedded PC, handheld computer, or a digital signal processor chip. Video recording can be triggered by the presence of motions and stored on a computer hard drive as AVI files. Motions can also automatically trigger an alarm or other security measures.

7.5 SMART ROADS

The same in-road detectors that control traffic lights could soon help unsnarl traffic jams, thanks to software developed by an Ohio State University engineer.

In tests, the software helped California road crews discover traffic jams up to three times faster, allowing them to clear accidents and restore traffic flow before drivers could be delayed. The technology could also be used to provide drivers with information for planning efficient routes or to improve future road designs, says Benjamin Coifman, assistant professor of electrical engineering and civil and environmental engineering at Ohio State University.

Many drivers have probably noticed the buried detectors, called loop detectors, at intersections. The devices are marked by the square outline that road crews create when they insert a loop of wire into the roadbed. When a car stops over the loop, a signal travels to a control box at the side of the road, which tells the traffic light to change. Although loop detectors are not much more than metal detectors, they collect enough information to indicate the general speed of traffic.

With Coifman's software, a small amount of roadside hardware, and a single PC, a city could use the detectors to significantly improve traffic monitoring without interfering with drivers. That's important, because good traffic management can't be obtrusive. "If transportation engineers are doing their job well, you don't even realize they've improved travel conditions," Coifman says.

Coifman's algorithms capture a vehicle's length as it passes over a detector. Once the car or truck passes over the next loop, the computer matches the two signals and calculates the vehicle's travel time. Based on the travel times, the software can spot emerging traffic jams within 3.5 minutes.

Because driver behavior isn't predictable, the algorithms take many human factors into account. Among other parameters, Coifman considered people changing lanes, entering and exiting from ramps, and "rubbernecking"—the delay to drive time caused by people who slow down to look at accidents or other events. "Traffic is a fluid like no other fluid," says Coifman. "You can think of cars as particles that act independently, and waves propagate through this fluid, moving with the flow or against it."

After an accident, it may take a long time for the telltale wave of slow-moving traffic to propagate through the detectors. With the new algorithm, Coifman can detect delays without waiting for slowed traffic to back up all the way to a detector. This improved response time is important, because personal and financial costs grow exponentially the longer people are stuck in traffic.

The detectors can't obtain any specific information about the make or model of a car, and a margin of error prevents the software from identifying more than a handful of cars in any one area at one time. But that's enough information to gauge traffic flow, and the benefits to motorists can be enormous. The average U.S. city dweller wastes 62 hours per year stuck in traffic, according to a 2002 urban mobility study published by the Texas Transportation Institute.

The Ohio Department of Transportation (ODOT) has already begun using loop detectors to help motorists spend less time in traffic. When drivers head south into Columbus on Interstate 71 during business hours, an electronic sign just north of the city displays the average drive time into downtown.

As such information becomes more common, drivers can plan their routes more efficiently, Coifman says. He's now working with ODOT to further improve travel time estimates. The software can work with other vehicle detection systems, such as video cameras. But installing these new systems can cost as much as $100,000 per location, yet retrofitting existing equipment to use Coifman's software would only cost a fraction as much.

7.6 CHIP IMPLANTS

Perhaps the final level of pervasive computing is human chip implants, which can be used to identify people or to track their movements. An implanted chip module could also give users the power to communicate and process information without relying on external devices. That will be good news for anyone who has misplaced a mobile phone, PDA, or laptop computer.

Some people favor chip implants because it would simplify life by eliminating the need to carry driver's licenses, passports, and other kinds of forgeable identification. Civil libertarians, on the other hand, shudder at the thought of giving governments and corporations the power to track and categorize people. Whether implantable chips ever become a widely used means of identification depends less on technology—for the basic technology is already here—and more on the attitudes of people and their governments.

7.6.1 Getting Under Your Skin

Global Positioning System (GPS) technology is popping up in a variety of products from PDAs to handheld navigation units to children's bracelets. But why lug around an external unit when you can have a GPS chip implanted inside your body? Applied Digital Solutions is working on just such a technology. The Palm Beach, Florida-based company has developed and recently concluded preliminary testing of a subdermal (under-the-skin) GPS personal location device (PLD).

A PLD could support various applications, including tracking missing children, hikers, or kidnapping victims. More ominously, repressive governments could use the technology to track the movements of political dissidents and other troublesome people.

The GPS chip includes a wireless receiver, transmitter, and antenna. Although satellite technology is used to determine a subject's location to within a few feet, the device must connect to a mobile phone network in order to relay information to outside parties.

The PLD prototype's dimensions are 2.5 inches in diameter by 0.5 inches in depth, roughly the size of a heart pacemaker. The company expects to be able to shrink the size of the device to at least one-half and perhaps to as little as one-tenth the current size. The device's induction-based power-recharging method is similar to the type used to recharge pacemakers. The recharging technique functions without requiring any physical connection between the power source and the implant.

Despite Applied Digital's recent progress, it's not likely that people will soon begin queuing up to receive GPS implants. "Implantable GPS is probably still at least five years away," says Ron Stearns, a senior analyst at Frost & Sullivan, a technology research firm located in San Antonio, Texas. Stearns notes that it will take that long to get a GPS chip down to the size where it can be implanted easily and comfortably. Applied Digital also faces regulatory hurdles, and a lengthy clinical trial process, before its chip can reach market. As a result, the technology is likely to be implanted into pets and livestock long before humans.

In the meantime, Applied Digital plans to continue testing its prototype to confirm that the device's transceiver, antenna, and power-recharging method are functioning properly. "We're very encouraged by the successful field testing and follow-up laboratory testing of this working PLD prototype," says Peter Zhou, Digital Solutions' vice president and chief scientist. "While reaching the working prototype stage represents a significant advancement in the development of [a] PLD, we continue to pursue further enhancements, especially with regard to miniaturization and the power supply. We should be able to reduce the size of the device dramatically before the end of this year."

7.6.2 Faster Fingerprints Via Wireless

New software will make it possible for law enforcement officials to capture, transmit, and process fingerprints anytime, anywhere. Atsonic, a Schaumburg, Illinois-based software developer, has introduced the first real-time wireless mobile fingerprint technology, designed for use by law enforcement, government agencies, security companies, the military, and even health care providers.

The company's SweetFINGER product is a wireless client-server technology that integrates biometrics, security standards, data management, and communications tools with proprietary real-time adaptive fingerprint search and identification capabilities. With SweetFINGER, fingerprint images can be scanned, compressed, and processed through a central system, transforming a procedure that often takes hours and even days to a few minutes.

The product aims to help law enforcement and security agents scan, collect, and verify an individual's fingerprints directly from crime scenes and check points or other field situations with a reliability and speed surpassing existing ink-based collection/verification methods. The software works with Panasonic's ToughBook01, a ruggedized handheld PC with built-in digitizer hardware.

State and local police must routinely check the identities of thousands of individuals using fingerprint data in different field locations, says Mugur Tolea, Atsonic's chief technical officer. "As a consequence, the data verification process is time and resource consuming." Checking a central databank, such as the FBI fingerprint database, can take hours or days, hampering law enforcement officials and wasting time that could be applied to crime investigation. "A real-time mobile device that can collect, transmit, and process fingerprints can prove to be an invaluable tool in strengthening law enforcement nationwide", says Tolea.

All biometric information collected with SweetFINGER is transferred securely to a law enforcement agency's or security organization's database server in Automated Fingerprint Identification System (AFIS) format. A complete profile of the individual identified, including demographic data, photo, and criminal background data, is transmitted back to the mobile device and displayed on its screen. The information can also be printed with SweetPRINT, an Atsonic proprietary mobile print solution. Besides collecting, transmitting, and processing fingerprints, SweetFINGER can also transfer photos of individuals via a digital camera attached to the handheld device.

SweetFINGER's scanning process is significantly faster than existing technologies—less than 0.1 second, according to Tolea. The scanned fingerprint has a 500 dpi resolution and is processed using more than 50 minutiae points. "It vastly improves the accuracy of biometrics data and speeds the time period for data collection, verification, and transmission, since it allows for real-time mobile processing of this information," says Joe Pirzadeh, Atsonic's president.

7.7 ENCRYPTION

The encryption applications market is receiving a huge boost from organizations' escalating security concerns due to increased use of the Internet for business-to-business transactions. "Increased research and development funding, venture capital, and investments in security infrastructure as well as new opportunities and applications are catapulting this market into a major growth area and attracting an increasing number of vendors eager to exploit these revenue opportunities," says James Smith, a security analyst at Technical Insights, a technology research firm located in San Jose, California.

In particular, public key infrastructure (PKI) technology, which provides a variety of crucial enabling capacities for electronic business processes, is expected to experience substantial revenue growth, as e-commerce becomes a mainstay of business-to-business transactions.

Virtual private networks (VPNs) and e-mail encryption applications also promise considerable opportunities to vendors. The ability to safely transfer sensitive data over the Internet, coupled with direct cost savings, will create increased demand for these technologies. However, with the cost factor acting as a major deterrent to many organizations, vendors will have to convince clients about the financial returns from securing networks to eliminate marketplace resistance and facilitate increased sales. Adds Smith, "Standard interoperability is critical to the success of the encryption applications market. Until this is achieved, market growth will be prevented from reaching its full potential."

In the VPN market, the Internet Protocol (IP) security standard is being helped by major projects such as the automotive network exchange extranet project. However, it remains unclear to what extent this standard has succeeded in ensuring interoperability. "Standardizing on IP security will enable multivendor interoperability that will allow the market to explode," says Smith.

Support of the secure multipurpose Internet mail extensions standard by major e-mail client and browser companies is helping to establish an approved standard in the e-mail encryption market.

In the PKI market, emerging standards such as on-line certificate status protocols, likely to gain recognition in two years, can help address multiple issues regarding certificate revocation and management of revocation on a large scale.

Spurring market growth further will be advanced forms of protection such as intrusion detection systems, tools that protect specific applications on corporate networks, and vulnerability-assessment software and services. Companies are expected to increasingly add intrusion detection systems to their security infrastructure to monitor inbound and outbound network activity.

7.7.1 A Double-Shot of Security Software

Network security software developed by researchers at Lucent Technologies' Bell Labs aims to make logging into network-based services and applications easier and more secure without sacrificing user privacy.

The security software consists of two complementary programs, called Factotum and Secure Store, which work together to prove a user's identity whenever he or she attempts to access a secure service or application, such as online banking or shopping. In contrast to systems that use a third party to control user information, Bell Labs' approach puts the user in control of personal information. Furthermore, Factotum and Secure Store are open platforms that can authenticate users with any Web site without requiring site operators to adopt any single sign-on standard.

Secure Store acts as a repository for an individual's personal information, whereas Factotum serves as an agent to handle authentication on the user's behalf in a quick, secure fashion. The approach tackles the problem of how to conveniently hold and use a diverse collection of personal information, such as usernames, passwords, and client certificates, for authenticating users to merchants or other services.

Factotum and Secure Store are inherently more secure because they give users control over their information, allow personal information to be stored on the network, not on a device, and use the latest protocols, says Al Aho, professor of computer science at Columbia University and the former Bell Labs vice president of computing sciences research. "Additionally, it's incredibly convenient because these applications eliminate the need for users to type the same information over and over, or to remember multiple passwords for each service they wish to access."

Although Factotum and Secure Store were both written for the Plan 9 operating system, an open-source relative of Unix developed at Bell Labs, they can be ported to other operating systems, including Linux, Windows, Solaris, and Unix.

"This technology has the potential to serve as the foundation for a new generation of more secure, easier-to-use authentication systems," says Eric Grosse, director of Bell Labs' networked computing research. "After using and improving Factotum and Secure Store in our own network and research lab, we are confident that they are ready for wider implementation."

To set up the Factotum and Secure Store services, a user first enters all of his or her various usernames and passwords into Secure Store. The network's Secure Store server protects this information using state-of-the art cryptography and the Advanced Encryption Standard (AES). To retrieve key files for Factotum, from a local device like a laptop or PDA, the user only needs to provide a password to prove his or her identity, thanks to PAK, an advanced Bell Labs security protocol for doing password-authenticated key exchange. This approach thwarts the most common security threats, like so-called "dictionary attacks" on the password, by making it impossible for someone to eavesdrop in on the challenge-and-response approach used in most password schemes.

When Factotum accesses a user's keys, it stores the information in protected memory and keeps it there for a short period of time. This is an improvement over today's common method of storing passwords on a user's hard drive, which is insecure. Factotum only holds user information in memory when the machine is running. When the machine is off, the secrets are only kept in Secure Store. The final security precaution is that Secure Store is located on the network, not on the user's PC; thus, even if a user's machine is hacked or stolen, the information stored in Secure Store is safe.

"The new security features in Plan 9 integrate organically into the system, making it unique among security options in the marketplace today," says David Nicol, professor of computer science at Dartmouth College and associate director of research and development at the school's Institute for Security Technology Studies. "Bell Labs' design recognizes rightly that identity and the authentication of identity are the heart and soul of security."

7.7.2 Data Hiding

Data hiding, the practice of secretly embedding data in images and other file types, promises to be a significant security concern in the years ahead. An electrical engineer at Washington University in St. Louis has recently devised a theory that sets the limits for the amount of data that can be hidden in a system and then provides guidelines for how to store data and decode it. Contrarily, the theory also provides guidelines for how an adversary would disrupt the hidden information. The theory is a fundamental and broad-reaching advance in information and communication systems that eventually will be imple mented in commerce and numerous homeland security applications—from detecting forgery to intercepting and interpreting messages sent between terrorists.

Using elements of game, communication, and optimization theories, Jody O'Sullivan, professor of electrical engineering at Washington University in St. Louis, and his former graduate student, Pierre Moulin, now at the University of Illinois, have determined the fundamental limits on the amount of information that can be reliably hidden in a broad class of data or information-hiding problems, whether they are in visual, audio, or print media. "This is the fundamental theorem of data hiding," O'Sullivan says. "One hundred years from now, if someone's trying to embed information in something else, they'll never be able to hide more than is determined by our theory. This is a constant. You basically plug in the parameters of the problem you are working and the theory predicts the limits."

Data, or information, hiding is an emerging area that encompasses such applications as copyright protection for digital media, watermarking, fingerprinting, steganography, and data embedding. Watermarking is a means of authenticating intellectual property—such as a photographer's picture or a Disney movie, by making imperceptible digital notations in the media identifying the owner. Steganography is the embedding of hidden messages in other messages. Data hiding has engaged the minds of the nation's top academics over the past seven years, but it has also caught the fancy of the truly evil. In February 2001, nine months before 9/11, USA Today reported that Osama bin Laden and his operatives were using steganography to send messages back and forth.

"The limit to how much data can be hidden in a system is key because it's important to know that you can't hide any more and if you are attacking (trying to disable the message) that you can't block anymore than this," O'Sullivan says. "It's also important because knowing this theory you can derive what are the properties of the optimal strategy to hide information and what are the properties of the optimal attack." O'Sullivan is associate director of Washington University's Center for Security Technologies, a research center devoted to developing technologies that safeguard the United States against terrorist attack.

Although the intellectual pursuit of data hiding is relatively new, with the first international conference on the topic held in 1996, the practice goes back to the ancient Greeks. Herodotus was known to have sent a slave with a message tattooed on his scalp to Mellitus; the slave grew his hair out to hide the message, which was an encouragement to revolt against the Persian king. In World War II, the Germans used microdots as periods at the ends of sentences. Magnified, the microdots carried lots of information. The German usage is a classic instance of steganography.

There will be much work ahead before O'Sullivan's theory will be fully implemented. "This is an example of one kind of work we do at the center that has a big impact in the theory community, but it's a couple of layers away from implementation," O'Sullivan says. "But the theory answers the questions, what is the optimal attack and what's the optimal strategy for information hiding."

7.7.3   Data Hiding's Positive Side

Although terrorists and other evil doers often use data hiding to disguise their nefarious plans, the technique has also a positive side. Copyright holders, for example, use data hiding to safeguard intellectual property, particularly images, sent over the Internet and various types of telecom devices.

Scientists from the University of Rochester and from Xerox have invented a new way to hide information within an ordinary digital image and to extract it again—without distortion of the original or loss of any information. Called "reversible data hiding," the new technique aims to solve a dilemma faced by digital image users, particularly in sensitive military, legal, and medical applications. Until now they have had to choose between an image that's been watermarked to establish its trustworthiness and one that isn't watermarked but preserves all the original information, allowing it to be enlarged or enhanced to show detail. When information is embedded with this newly discovered method, authorized users can do both.

"Commonly used techniques for embedding messages such as digital watermarking irreversibly change the image, resulting in distortions or information loss. While these distortions are often imperceptible or tolerable in normal applications, if the image is enlarged, enhanced, or processed using a computer, the information loss can be unacceptable," says Gaurav Sharma, an imaging scientist at Xerox's Solutions and Services Technology Center in Webster, New York.

"With our new data embedding algorithm, authorized recipients not only can extract the embedded message but also can recover the original image intact, identical bit for bit to the image before the data was added," he says. "The technique offers a significantly higher capacity for embedding data and/or a lower-distortion than any of the alternatives."

"The technique will be widely applicable to situations requiring authentication of images with detection of changes, and it can also be used to encode information about the image itself, such as who took the picture, when or with what camera," adds Murat Tekalp, professor of electrical engineering at the University of Rochester. "The greatest benefit of this technology is in determining if anyone has clandestinely altered an image. These days many commercial software systems can be used to manipulate digital images. By encoding data in this way, we can be sure the image has not been tampered with, and then remove the data within it without harming the quality of the picture," he says.

Although the technique is currently implemented in software, implementation in hardware or firmware in trusted devices where image integrity is critical to the application could be possible. For instance, the technique could be used in a trusted digital camera used to gather forensic evidence to be later used at a trial. If information is embedded in the images captured with the camera using the new algorithms, any subsequent manipulations of the pictures could be detected and the area where they occurred pinpointed.

7.8 QUANTUM CRYTOGRAPHY

Quantum cryptography is a codemaker's Holy Grail. The idea is to use a rapid series of light pulses (photons) in one of two different states to transmit information in an unbreakable code. Quantum cryptography differs from other code schemes in that the attempt by a third party to intercept a code's key itself alters the key. It is as though the very act of listening in on a conversation makes the eavesdropper known.

Let us say that Bill wants to send a secret message to Judie. For the message to be secret, Bill has to employ some scheme to encode the message. And Judie needs a key to the scheme to decode the message. The crucial communication for the sake of preserving secrecy is not the message but the key. Therefore, Bill sends a string of single photons whose polarizations successively contain the key. If a third party, John, tries to detect the singly transmitted photons, the act of detection causes an irreversible change in the wave function of the system. ("Wave function" denotes the quantum mechanical state of a physical system.) If John then tries to send the key on to Judie, the key will, in effect, bear the imprint of John's intermediate detection.

7.8.1 Quantum Dots

Utilizing the power of quantum cryptography, a National Institute of Standards and Technology (NIST) scientist has demonstrated efficient production of single photons—the smallest pulses of light—at the highest temperatures reported for the photon source used. This advance marks a step toward practical, ultrasecure quantum communications.

"Single photon turnstiles" are being hotly pursued for quantum communications and cryptography, which involve the use of streams of individual photons in different quantum states to transmit encoded information. Due to the peculiarities of quantum mechanics, such transmissions could not be intercepted without being altered, thus ensuring that eavesdropping would be detected.

The photon source used in the NIST study was a "quantum dot," 10 to 20 nm wide, made of semiconductor materials. Quantum dots have special electronic properties that, when excited, cause the emission of light at a single wavelength that depends on dot size. An infrared laser tuned to a particular wavelength and intensity was used to excite the quantum dot, which produced photons one by one more than 91 percent of the time at temperatures close to absolute zero (5 K or about minus 459 degrees F) and continued to work at 53 percent efficiency at 120 K (minus 243 degrees F). Higher operating temperatures are preferable from a cost standpoint because the need for cooling is reduced.

The NIST quantum dots are made of indium gallium arsenide, can be fabricated easily, and can be integrated with microcavities, which increase photon capture efficiency. According to NIST electrical engineer Richard Mirin, this design offers advantages over other single photon sources, many of which exhibit blinking, stop working under prolonged exposure to light, or are difficult to fabricate.

7.8.2 Quantum Photon Detector

Steps are also being made to transform quantum encryption from a theoretical possibility into a practical technology. Researchers from NIST and Boston University have demonstrated a detector that counts single pulses of light, while simultaneously reducing false or "dark counts" to virtually zero. The advance provides a key technology needed for future development of secure quantum communications and cryptography.

Most current photon detectors operate best with visible light, cannot reliably detect single photons, and suffer from high dark counts due to random electronic noise. The new device operates with the wavelength of near-infrared light used for fiber optic communications and produces negligible dark counts. Instead of using light-sensitive materials, the NIST device uses a tungsten film coupled to a fiber-optic communication line. The film is chilled to minus 120,000,000, at its transition temperature between normal conductivity and superconductivity. When the fiber-optic line delivers a photon to the tungsten film, the temperature rises and the apparatus detects it as an increase in electrical resistance. The device detects about 20,000 photons per second and works with an efficiency of about 20 percent. With planned improvements, the research team hopes to increase efficiencies to greater than 80 percent.

7.8.3 Distance Record

Although quantum encryption sounds good in theory, the technique must be able to cope with real-word conditions. Specifically, the approach must be compatible with everyday communications networks. Researchers at Toshiba Research Europe have broken the distance record for the only potentially hacker-proof form of communications: quantum cryptography.

The Toshiba research team, based in Cambridge, United Kingdom, was the first in the world to demonstrate successful quantum cryptography over 100 kilometers of optical fiber, showing the possibility of broad commercial potential. Likely quantum cryptography adopters include any organization that uses Internet and communications technologies to send, receive, and store sensitive information, including banks, retailers, and central and local government organizations.

Cryptography, the science of information security, is essential to protecting electronic business communication and e-commerce transactions. It allows message confidentiality, user identification, and secure transaction validation. Much of the interest in quantum cryptography stems from the fact that the approach is fundamentally secure. This contrasts with today's code-based systems, which rely on the assumed difficulty of certain mathematical operations. Quantum cryptography would provide a communication method utilizing secrecy that doesn't depend on any assumptions.

Quantum cryptography allows two users on an optical fiber network to form a shared key, the secrecy of which can be guaranteed. The technology takes advantage of the particle-like nature of light. In quantum cryptography, each transmitted bit is encoded on a single light particle (or photon). The impossibility of faithfully copying this stream of encoded photons ensures that a hacker can never determine the key without leaving detectable traces of intervention.

Until now, quantum cryptography's major limitation is that light particles could be scattered out of the fiber. In theory, this isn't critical, since only a tiny fraction of the photons that reach a fiber's end point are used to form the key. In practice, however, the rate of photons surviving long fibers can be so low that they are masked by noise in the photon detector. By developing an ultra-low noise detector, the Toshiba team has been able to demonstrate a system working over much longer fibers than achieved previously.

"As far as we are aware, this is the first demonstration of quantum cryptography over fibers longer than 100 kilometers," says Andrew Shields, who leads the Toshiba group developing the system. "These developments show that the technique could be deployed in a wide range of commercial situations within a time frame of less than three years."

Michael Pepper, joint managing director of Toshiba Research Europe, notes that advancements in semiconductor technology are allowing his researchers to implement quantum effects that were previously known only in theory. "One can foresee that this is the beginning of a process which will lead to a revolution in information processing and transmission," he says.

7.9 E-MAIL "CLUSTER BOMBS"

The biggest security threat to PC-based telecom devices, such as laptop computers, is e-mail-delivered viruses, worms, and other types of malicious codes. Now it's time to prepare for a fresh Internet scourge—e-mail "cluster bombs." This new online threat, which inundates user e-mail in-boxes with hundreds or thousands of messages in a short period of time, promises to paralyze PDAs and other Internet access devices, says a computer researcher.

According to Filippo Menczer, an associate professor of informatics and computer science at Indiana University at Bloomington, a weakness in Web site design makes e-mail cluster bombs possible. The technique is typically employed to target an individual. A miscreant poses as the victim and, using the victim's own e-mail address, fills out Web site forms, such as those used to subscribe to a mailing list.

One or two automated messages would hardly overload an e-mail inbox. But software agents, Web crawlers, and scripts can be used by the bomber to fill in thousands of forms almost simultaneously, resulting in a "cluster bomb" of unwanted automatic reply e-mail messages to the victim. The attack can also target a victim's mobile phone with a sudden, large volume of Short Message Service (SMS) messages.

"This is a potential danger but also a problem that is easy to fix," says Menczer. "We wanted to let people know how to correct the problem before a hacker or malicious person exploits this vulnerability, causing real damage." The barrage of messages would dominate the bandwidth of an Internet connection, making it difficult or impossible for the victim to access the Internet. This is called a distributed denial-of-service attack because a large number of Web sites attack a single target.

The attack works because most Web forms do not verify the identity of the people—or automated software—filling them out. But Menczer says there are some simple things Web site managers can do to prevent attacks. "Often, subscribing to a Web site results in an automatically generated e-mail message asking the subscriber something like, 'Do you want to subscribe to our Web site?'" he notes. "We propose that Web forms be written so that the forms do not cause a message to be sent to subscribers at all. Instead, the form would prompt subscribers to send their own e-mails confirming their interest in subscribing. This would prevent the Web site from being abused in a cluster bomb attack."

Menczer conducted his research with Markus Jakobsson, principal Research Scientist at RSA Laboratories, a computer security firm, located in Bedford, Mass. Funding for the study came from an National Science Foundation Career Grant and the Center for Discrete Mathematics and Theoretical Computer Science at Rutgers University.

Chapter 8

Energy to Go— Power Generation

Given the number of portable electronic devices flooding the market, the need for cheap, portable, and quickly renewable electrical power sources has never been greater. Yet, despite researchers' best efforts, today's portable power sources remain stubbornly bulky and expensive.

8.1 NEW MATERIALS

New materials promise to extend the performance of conventional battery technologies while opening the door to new types of power sources. Scientists at the U.S. Department of Energy's Sandia National Laboratories in Livermore, California, are among the researchers investigating promising new power-storage substances. The scientists' initial work appears promising. Sandia's researchers have developed a new class of composite anode materials—composed of silicon and graphite—that may double the energy storage capacities presently possessed by graphite anodes. The anode is the negative electrode—or battery area—where electrons are lost. The breakthrough could lead to rechargeable lithium ion batteries with more power, longer life, and smaller size.

The marriage of silicon and graphite may improve the capabilities of commercial graphite anode materials by up to 400 percent, says Jim Wang, an analytical materials science manager at Sandia. "Electronics designers are

Telecosmos: The Next Great Telecom Revolution, edited by John Edwards
ISBN 0-471-65533-3 Copyright © 2005 by John Wiley & Sons, Inc.

currently forced to use low-power-consumption components and designs that are limited in their longevity," he says. "Our newly discovered anode materials can improve the performance of microsystems by allowing for more powerful, sophisticated electronic components and by reducing the size and weight of the overall system."

For years, researchers have been vexed by the capacity limits associated with traditional lithium battery anodes. As a result, Sandia turned to silicon, which offers more than 10 times the lithium capacity potential of graphite but is hampered itself by a rapid capacity loss over time. When small particles of silicon are combined within a graphite matrix, however, large capacities can be retained over time. "The promising aspects of these materials are the large capacities, the capacity retention during cycling compared to other high-capacity materials, and the ability to control its performance by changing the composite composition and microstructure," says Wang.

Karl Gross, one of the principal investigators on the Sandia research team, says the silicon-graphite material could be produced via a simple milling process. The production technique is commonplace in the battery industry. In addition, the raw materials needed to produce the material are inexpensive and abundant.

The Sandia researchers admit, however, that their new material has some potential vulnerabilities. The biggest obstacle is that complete elimination of long-term capacity fading may not be possible, although it can likely be minimized by the design of the silicon-graphite composite structure. Yet Wang is confident that silicon-graphite electrodes will set the bar for future breakthroughs. "We believe that only other silicon-containing electrode materials can compete with the large capacities that our silicon-graphite composites have demonstrated," he says.

"Manufacturers of electric automobiles, laptop computers, cell phones, power tools, and other hybrid microsystems will likely all benefit from this kind of technology," says Scott Vaupen, a spokesman for Sandia's business development department. Vaupen notes Sandia hope to collaborate with others to further develop the technology for eventual licensing and commercialization.

8.1.1 Carbon Nanotube Batteries

Experiments with carbon nanotubes, a new form of carbon discovered about a decade ago, have recently suggested that it should be possible to store more energy in batteries using the tiny tubes than with conventional graphite electrodes. The experiments, conducted at the University of North Carolina at Chapel Hill, show carbon nanotubes can contain roughly twice the energy density of graphite. One possibility, researchers say, is longer-lasting batteries.

"Scientists and others, including the popular press, have shown a lot of interest in carbon nanotubes because of numerous potential applications," says Zhou. "They are very strong tubular structures formed from a single layer of carbon atoms and are only about a billionth of a meter in diameter." Beyond batteries, uses may include constructions of flat-panel displays, telecom devices, fuel cells, high-strength composite materials, and novel molecular electronics.

The experiments have demonstrated for the first time that nanotubes work better than conventional materials. "In our experiments, we used both electrochemistry and solid state nuclear magnetic resonance measurements, which show similar results," says Zhou. "With graphite, we can store, reversibly, one charged lithium ion for every six carbon atoms in graphite, but we found that with nanotubes, we can store one charged lithium ion for every three carbons, also reversibly."

Most rechargeable batteries in portable electronics today are lithium ion batteries, which use graphite or carbonaceous materials as one of the electrodes. Reactions occurring at the electrodes create a flow of electrons that generate and store energy. The UNC scientists created single-wall carbon nanotubes by subjecting a carbon target to intense laser beams. By chemical processing, the researchers were able to open the closed ends of the nanotubes and reduce their lengths. "This allows the diffusion of lithium ions into the interior space of the nanotubes and reduces the diffusion time," Zhou says. "We believe this is the reason for the enhanced storage capacity."

Better batteries are possible because the team found a significantly higher energy density, he says. "We have shown this for the first time experimentally," says Zhou. "Now, we'll have to work on and overcome other practical issues before we can make real devices, but we are very optimistic."

8.1.2 Thin Films

Research being conducted at Ohio State University could lead to PDAs and other mobile devices featuring enhanced displays and the ability to run off of solar power.

Steven Ringel, an Ohio State professor of electrical engineering, and his colleagues have created unique hybrid materials that are virtually defect-free—an important first step for making ultra-efficient electronics. The same technology could also lead to faster, less expensive computer chips.

Ringel directs Ohio State's Electronic Materials and Devices Laboratory, where he and his staff grow thin films of "III–V" semiconductors—materials made from elements such as gallium and arsenic, which reside in groups III and V of the chemical periodic table. Because III–V materials absorb and emit light much more efficiently than silicon, these materials could bridge the gap between traditional silicon computer chips and light-related technologies, such as lasers, displays, and fiber optics.

Researchers have tried for years to combine III–V materials with silicon but have achieved only limited success. Now that Ringel has succeeded in producing the combination with high quality, he has set his sights on a larger goal.

"Ultimately, we'd like to develop materials that will let us integrate many different technologies on a single platform," Ringel says.

The key to Ringel's strategy is the idea of a "virtual substrate"—a generic chip-like surface that would be compatible with many different kinds of technologies and could easily be tailored to suit different applications. Ringel's current design consists of a substrate of silicon topped with III–V materials such as gallium and arsenide, with hybrid silicon-germanium layers sandwiched in between. The substrate is 0.7 mm thick, whereas the gallium arsenide layer is only 3 mm—millionths of 1 meter—thick.

Other labs have experimented with III–V materials grown on silicon, but none has been able to reduce defect levels below a critical level that would enable devices like light-emitting diodes and solar cells to be achieved, Ringel says. Defects occur when the thin layers of atoms in a film aren't lined up properly. Small mismatches between layers rob the material of its ability to transmit electrical charge efficiently. Ringel and his colleagues grew films of III–V semiconductors with a technique known as molecular beam epitaxy, in which evaporated molecules of a substance settle in thin layers on the surface of the silicon-germanium alloy. To search for defects, they used techniques such as transmission electron microscopy.

Defects are missing or misplaced atoms that trap electrons within the material, Ringel explains. That's why engineers typically measure the quality of a solar cell material in terms of carrier lifetime—the length of time an electron can travel freely through a material without falling into a defect. Other experimental III–V materials grown on silicon have achieved carrier lifetimes of about two nanoseconds, or two billionths of a second. Ringel's materials have achieved carrier lifetimes in excess of 10 nanoseconds.

The engineers have crafted the III–V material into one-square-inch versions of solar cells in the laboratory and achieved 17 percent efficiency at converting light to electricity. They have also built bright light-emitting diodes (LEDs) on silicon substrates that have a display quality comparable to that of traditional LEDs.

8.2 SMALLER, LIGHTER POWER ADAPTER

As notebook computers become thinner and lighter, the ubiquitous power adapter remains stubbornly heavy and bulky. In fact, it's not hard to see the day when notebooks are actually more compact than their power adaptors. But smaller and lighter adapters may soon be on the way, thanks to a little known power-generation approach known as piezoelectric technology.

Transformers are needed to convert the 115-volt, 60-cycle power available from a standard U.S. wall receptacle to the 13 to 14 volts direct current used by laptops. "Electromagnetic transformers are shrinking slightly, but there are theoretical limitations in reducing the general size," says Kenji Uchino, a professor of electrical engineering at Pennsylvania State University. However, a piezoelectric motor and transformer can be much smaller and lighter, he notes.

Piezoelectric material moves when placed under an electric voltage. Additionally, when displaced by outside pressure, piezoelectric materials produce an electric voltage. Transformers are made from piezoelectric materials by applying a chopped electric voltage to one side of a piezoelectric wafer. This on and off voltage creates a vibration in the material, which is converted to an AC voltage on the other side of the wafer. The amount of increase or decrease in the voltage transformed is dependent on the gap between the electrodes.

Most laptops require about 15 volts direct current with less than 1 amp of current and about 12 watts of power. By manipulating the length and width of the piezoelectric chip, the researchers can convert 115 volts to 15 volts. A rectifier then converts the alternating current to direct current. "Smaller, less complex piezoelectric devices are already in use as step-up transformers in some laptops to light the monitors, which can take 700 volts to turn on and 50 to 150 volts to continue their operations," says Uchino.

A key advantage of piezoelectric PC power adapters is that they do not produce the heat that conventional electromagnetic transformers produce. Electromagnetic power adapters also produce heat, noise, and interference. Piezoelectric power adapters operate in the ultrasonic range, so humans cannot hear any sound produced, and they do not produce any electromagnetic interference.

"Right now, we can reduce the adapters to one-fourth of their current size," says Uchino. "Eventually, we would like to make it the size of a pen, but that is far away." Uchino wants to make an adaptor that's small enough to become an internal part of the notebook, eliminating the need for users to carry a separate "power brick" on their journeys. Uchino notes that, although his group is targeting the notebook computer market, small piezoelectric adapters are suitable for any appliance that requires an AC to DC converter and transformer.

The Penn State researchers are developing the technology in collaboration with Virginia-based Face Electronics and Taihciyo Cement Corporation of Japan.

8.2.1 Glass Battery

Another material that shows power storage promise is glass—the same material used in everything from car windshields to beer bottles. As far as inventions go, a glass battery sounds about as promising as a concrete basketball or an oatmeal telephone. But inventor Roy Baldwin claims that his unique power source could someday energize everything from mobile phones to automobiles.

Baldwin's battery is based on Dynaglass, an inorganic polymer that's allegedly stronger than steel, yet flexible enough to wrap food. Baldwin, a retired mechanical engineer, says Dynaglass was developed in the mid-1990s, but some of the technology is based on research by the Soviet military and space programs. He learned about the material while helping a friend ship medical supplies to Russia. "Later on, we discovered that the material could be used to store energy," says Baldwin, who then formed a company—Columbus, Ohio-based Dynelec—to explore the technology's potential.

Baldwin boasts that Dynaglass is a remarkable power source. A Dynaglass battery, he says, is infinitely rechargeable and might be able to generate up to 30 times more energy than a lead-acid battery of comparable size and weight. The device, which can be produced in a wide range of sizes, also contains no acids or other dangerous chemicals, making it pollution free. "It just reacts like glass," Baldwin says. But since Dynaglass isn't brittle like ordinary glass, it's durable and won't shatter when dropped.

While a working Dynaglass battery would be warmly received by mobile device manufacturers and users, Keith Keefer, a scientist based in Richland, Washington, is skeptical that Baldwin's technology is all that it's purported to be. He notes that several inventors have created similar devices, and that none of the devices has lived up to its promise. "No one has ever really made it work," he says.

Yet Baldwin is looking to interest manufacturers in his technology. "A glass producer could use this to enter the energy industry at practically no additional cost," he says.

8.2.2 Ion Trek

Mobile phones, CD players, and flashlights all wear down batteries far faster than we might wish. Researchers at the U.S. Department of Energy's Idaho National Engineering and Environmental Laboratory, however, have overcome another barrier to building more powerful, longer-lasting lithium-based batteries. The team, led by inorganic chemist Thomas Luther, has discovered how lithium ions move through the flexible membrane that powers their patented rechargeable lithium battery.

Luther describes the translucent polymer membrane as an "inorganic version of plastic kitchen wrap." The team, including chemists Luther, Mason Harrup, and Fred Stewart, created it by adding a ceramic powder to a material called MEEP ([bis(methoxyethoxyethoxy)phosphazene]), an oozy, thick oil. The resulting solid, pliable membrane lets positively charged lithium ions pass through to create the electrical circuit that powers the battery but rebuffs negatively charged electrons. This keeps the battery from running down while it sits on the shelf—overcoming a major battery-life storage problem.

For years, rechargeable lithium battery performance has been disappointing because the batteries needed recharging every few days. After conquering the discharge challenge, the team attacked the need for greater battery power to be commercially competitive. Their membrane didn't allow sufficient passage of lithium ions to produce enough power, so they needed to understand exactly how the lithium ions move through the membrane on a molecular level. First, they analyzed the MEEP membrane using nuclear magnetic resonance—the equivalent of a hospital MRI—to zero in on the best lithium ion travel routes. The results supported the team's suspicion that the lithium ions travel along the membrane's "backbone." The MEEP membrane has a backbone of alternating phosphorus and nitrogen molecules, with oxygen-laden "ribs" attached to the phosphorus molecules.

Further analysis with infrared and raman spectroscopy (techniques that measure vibrational frequencies and the bonds between different nuclei) helped confirm that the lithium ions are most mobile when interacting with nitrogen. Lithium prefers to nestle into a "pocket" created by a nitrogen molecule on the bottom with oxygen molecules from a MEEP rib on either side. Armed with this new understanding of how lithium moves through the solid MEEP membrane, the team starting making new membrane versions to optimize lithium ion flow. This should make the team's lithium batteries much more powerful. The team's research results are a major departure from the conventionally accepted explanation of lithium ion transport that proposed the lithium/MEEP transport mechanism as jumping from one rib to the next using the oxygen molecules as stepping stones.

Harrup, Stewart, and Luther are optimistic their battery design will ultimately change the battery industry. The team projects that its polymer membrane will be so efficient at preventing battery run down, that batteries could sit unused for up to 500 months between charges with no loss of charge. Because the membrane is a flexible solid, it can be molded into any shape, which could open up new applications for batteries. The membrane is also very temperature tolerant, which could potentially solve portable power need problems in the frigid cold of space. The team is already working with several federal agencies on applications for its lithium battery designs.

8.3 FUEL CELLS

Although PC vendors are eager to breathe new life into their aging systems, particularly modestly equipped notebooks, at least one highly anticipated technology may not make it into the mainstream as soon as many vendors would like.

Micro fuel cell technology has been aggressively touted as a convenient and easily renewable power source. The devices, which generate electricity through a chemical reaction between oxygen and a fuel such as hydrogen or methanol, can power a notebook for up to 40 hours. Yet, it's unlikely that large numbers of users will be "filling up" notebook PCs, PDAs, and other mobile devices anytime soon. Roadblocks for use include fuel cell size, the lack of a universal standard, customer education issues, and safety and security concerns as users bring devices containing volatile fluids into buildings and onto airplanes and other vehicles.

All of these drawbacks have made many notebook vendors skeptical about fuel cell technology. "Fuel cells are not likely to be relevant for mainstream mobile devices for several years," says Jay Parker, notebook products director for Round Rock, Texas-based Dell Computer. He believes it will be hard to change notebook users' ingrained habits. "Customers will need to become acclimated to refueling rather than recharging." Parker notes, however, that Dell is continuing to evaluate various fuel cell technologies.

Howard Locker, chief architect of Armonk, New York-based IBM's personal computing division, says fuel cells will never become popular because users will have to pay for each refill. "Today, when you charge a battery, it's free," he says. "Folks are already at nine hours on a battery, so how much better does it need to get?" Locker's opinion of fuel cell technology: "It's a nonstarter."

Two notebook makers, however, are undeterred by the naysayers and plan to push ahead with fuel cell technology. Toshiba and NEC have each announced they will start selling fuel cell-equipped notebooks during 2004.

8.4 MICROCOMBUSTION BATTERY

The search for a better battery is getting a push from the U.S. Defense Advanced Research Project Agency (DARPA), which has given Yale University's engineering department $2.4 million to develop readily rechargeable microcombustion batteries.

The Yale research is part of DARPA's Palm Power program, which addresses the military's need for lighter and more compact electrical power sources. "DARPA is shooting for something that weighs as little as a few ounces to power the growing number of communications and weapons systems that tomorrow's soldiers will carry," says Alessandro Gomez, director of the Yale Center for Combustion Studies and a professor of mechanical engineering.

Microcombustion technology generates heat by slowly burning tiny amounts of liquid hydrocarbons. The heat is then converted into electricity by other energy conversion schemes such as thermoelectric and thermophotovoltaic. By taking advantage of the abundant power densities offered by hydrocarbon fuels, a microcombustion battery with millimeter-level dimensions could provide the same power and operating time as a conventional battery up to 10 times its size. And microcombustion cells could be quickly refueled with an eyedropper.

The Department of Defense plans to use microcombustion batteries in everything from tactical bodyware computers to Micro Air Vehicles—six-inch-long unmanned reconnaissance aircraft. The technology, once perfected, should spill over quickly into business and consumer products, Gomez says. "Laptop computers, cell phones, and a variety of other portable electronics products could all benefit."

Yale scientists are concentrating on developing the most effective combustion technology, while researchers at other institutions are working on techniques for converting thermal energy into electrical energy. "Conventional battery technology has reached a dead end," says Gomez. "We're looking to develop a power source that's every bit as innovative as the latest military systems."

8.5   POWER MONITOR

As people increasingly rely on sophisticated mobile phones and PDAs to handle an array of tasks, knowing exactly how much battery power remains inside a device becomes ever more critical, especially before accessing important information or initiating a wireless transaction.

Texas Instruments is looking to help mobile device users accurately monitor their power usage. The company has developed the first fully integrated battery fuel gauge for single-cell lithium ion and lithium polymer battery packs. The chip-based gauge is designed to help users observe remaining battery capacity and system run time (time to empty).

The chip, named bqJunior, promises to help manufacturers reduce the development time and total cost of implementing a comprehensive battery fuel gauge system in mobile devices. "As battery-powered consumer devices become more complicated and dynamic, designers of those products will require the right intelligent hardware to provide accurate information about the battery and system run times to better manage available power," says Peter Fundaro, worldwide marketing manager for Texas Instrument's battery management products. Fundaro notes that Texas Instrument's product simplifies the design of a cost-effective accurate battery fuel gauge in single cell by "offering a solution that performs all the necessary intelligent calculations on-chip, significantly reducing the amount of calculations performed by the host-side microcontroller."

Unlike a standard battery monitor, bqJunior incorporates an on-board processor to calculate the remaining battery capacity and system run-time. The device measures the battery's charge and discharge currents to within 1 percent error using an integrated low-offset voltage-to-frequency converter. An analog-to-digital converter measures battery voltage and temperature. Using the measurement inputs, the bqJunior runs an algorithm to accurately calculate remaining battery capacity and system run time.

bqJunior compensates remaining battery capacity and run times for battery discharge rate and temperature variations. Because the device performs the algorithm and data set calculations, there's no need to develop and incorporate code to implement those tasks in the host system processor, which helps reduce development time and total implementation cost. The host system processor simply reads the data set in bqJunior to retrieve remaining battery capacity, run time, and other critical information that's fundamental to comprehensive battery and power management, including available power, average current, temperature, voltage, and time to empty and full charge. bqJunior includes a single wire communications port to communicate the data set to the system host controller. The fuel gauge operates directly from a single lithium ion cell and operates at less than 100 microamps. It features three low-power standby modes to minimize battery consumption during periods of system inactivity.

Other bqJunior features include a low-offset voltage-to-frequency converter (VFC) for accurate charge and discharge counting. Also provided are an internal time base and an on-chip configuration EEPROM that allows application-specific parameters. bqJunior takes advantage of Texas Instrument's new LBC4 copper CMOS process node, which helps achieve higher integration, lower power, and enhanced performance.

8.6 COOLING TECHNOLOGIES

Two new technologies developed at the Georgia Institute of Technology promise to remove heat from electronic devices and could help future generations of laptops, PDAs, mobile phones, telecom switches, and high-powered military equipment keep their cool in the face of growing power demands.

The technologies—synthetic jet ejector arrays (SynJets) and vibration-induced droplet atomization (VIDA)—are designed to keep telecom devices cool despite relentless miniaturization. "There is a lot of concern in the electronics industry about thermal management," says Raghav Mahalingam, a Georgia Tech research engineer and the technology's codeveloper. "New processors are consuming more power, circuit densities are getting higher, and there is pressure to reduce the size of devices. Unless there is a breakthrough in low-power systems, conventional fan-driven cooling will no longer be enough."

Processors, memory chips, graphics chips, batteries, radio frequency components, and other devices found in electronic equipment generate heat that must be dissipated to avoid damage. Traditional cooling techniques use metallic heat sinks to conduct thermal energy away from the devices, then transfer it to the air via fans. However, cooling fans have a number of limitations. For instance, much of the circulated air bypasses the heat sinks and doesn't mix well with the thermal boundary layer that forms on the fins. Fans placed directly over heat sinks have "dead areas" where their motor assemblies block airflow. Additionally, as designers boost airflow to increase cooling, fans use more energy, create more audible noise, and take up more space.

8.6.1 SynJets

Developed by Mahalingam and Ari Glezer, a professor at Georgia Tech's School of Mechanical Engineering, SynJets are more efficient than fans, producing two to three times as much cooling with two-thirds less energy input. Simple and with no friction parts to wear out, a synthetic jet module in principle resembles a tiny stereo speaker in which a diaphragm is mounted within a cavity that has one or more orifices. Electromagnetic or piezoelectric drivers cause the diaphragm to vibrate 100 to 200 times per second, sucking surrounding air into the cavity and then expelling it. The rapid cycling of air into and out of the module creates pulsating jets that can be directed to the precise locations where cooling is needed.

The jet cooling modules take up less space in cramped equipment housings and can be flexibly conformed to components that need cooling—even mounted directly within the cooling fins of heat sinks. Arrays of jets would provide cooling matched to component needs, and the devices could even be switched on and off to meet changing thermal demands. Although the jets move 70 percent less air than fans of comparable size, the airflow they produce contains tiny vortices that make the flow turbulent, encouraging efficient mixing with ambient air and breaking up thermal boundary layers.

"You get a much higher heat transfer coefficient with synthetic jets, so you do away with the major cooling bottleneck seen in conventional systems," says Mahalingam. The ability to scale the jet modules to suit specific applications and to integrate them into electronic equipment could provide cooling solutions over a broad range of electronic hardware ranging from desktop computers to PDAs, mobile phones, and other portable devices that are now too small or have too little power for active cooling.

SynJets could be used by themselves to supplement fans or even in conjunction with cooling liquid atomization. "We will fit in where there currently is no solution or improve on an existing solution," says Jonathan Goldman, a commercialization catalyst with Georgia Tech's VentureLab, a program that helps faculty members commercialize the technology they develop. Beyond the diaphragm, the system requires an electronic driver and wiring. Goldman expects the jets to be cost competitive with fans and easier to manufacture. Further energy savings could be realized by using piezoelectric actuators.

One of the practical implications of this technology could be to forestall the need to use costlier heat sinks made from copper. "The industry could continue to use aluminum and retain its advantages of design simplicity, lower cost, and lower weight," says Goldman.

8.6.2   VIDA

In applications like high-powered military electronics, automotive components, radars, and lasers, power dissipation needs exceed 100 watts per square centimeter and may surpass 1,000 watts per square centimeter. For such higher demands, vibration-induced droplet atomization (VIDA) could be used. This sophisticated system uses atomized liquid coolants—such as water—to carry heat away from components. Also developed at Georgia Tech by Glezer's group, VIDA uses high-frequency vibration produced by piezoelectric actuators to create sprays of tiny cooling liquid droplets inside a closed cell attached to an electronic component in need of cooling.

The droplets form a thin film on the heated surface, allowing thermal energy to be removed by evaporation. The heated vapor then condenses, either on the exterior walls of the cooling cell or on tubes carrying liquid coolant through the cell. The liquid is then pumped back to the vibrating diaphragm for reuse. "A system like this could work in the avionics bay of an aircraft," says Samuel Heffington, a Georgia Tech research engineer. "We have so far been able to cool about 420 watts per square centimeter and ultimately expect to increase that to 1,000 watts per square centimeter."

SynJets and VIDA have both been licensed to Atlanta-based Innovative Fluidics, which will use them to develop products that will be designed to meet a broad range of electronic device cooling needs.

8.6.3 Wiggling Fans

Another promising approach to device cooling is based on tiny, quiet fans that wiggle back and forth to help cool future laptop computers, mobile phones, and other portable electronic gear.

The devices, developed by researchers at Purdue University, aim to remove heat by waving a small blade in alternate directions, like the motion of a classic hand-held Chinese fan (Fig. 8-1). They consume only about 1/150th as much

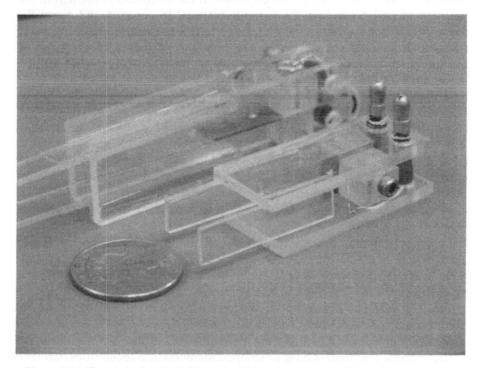

Figure 8-1 *Tiny, quiet fan that will help cool future laptop computers, mobile phones and other portable electronic gear.*

electricity as conventional fans, and they have no gears or bearings, which produce friction and heat. Because the new fans work without motors that contain magnets, they do not produce electromagnetic "noise" that can interfere with electronic signals in computer circuits.

The cramped interiors of laptop computers and cell phones contain empty spaces that are too small to house conventional fans but large enough to accommodate the new fans, some of which have blades about an inch long. Placing the fans in these previously empty spaces has been shown to dramatically reduce the interior temperatures of laptop computers. The wiggling fans will not replace conventional fans. Instead, they will be used to enhance the cooling now provided by conventional fans and passive design features, such as heat-dissipating fins.

In experiments on laptop computers, the Purdue researchers reduced the interior temperatures by as much as 8 degrees Celsius. "For a very small power expenditure, we are able to get a huge benefit," says Suresh Garimella, an associate professor of mechanical engineering at Purdue. The fans run on 2 milliwatts of electricity, or 2 1/1,000ths of 1 watt, compared with 300 milliwatts for conventional fans.

The fans are moved back and forth by a "piezoelectric" ceramic material that is attached to the blade. As electricity is applied to the ceramic, it expands, causing the blade to move in one direction. Then, electricity is applied in the alternate direction, causing the ceramic material to contract and move the blade back in the opposite direction. This alternating current causes the fan to move back and forth continuously. The operating efficiency of a fan can be optimized by carefully adjusting the frequency of alternating current until it is just right for that particular fan.

The piezoelectric fans can be made in a wide range of sizes. The Purdue engineers are developing fans small enough to fit on a computer chip: their blades will only be about 100 micrometers long, which is roughly the width of a human hair.

Piezoelectric fans were developed during the 1970s. The first versions were considered noisy, but the Purdue group has developed fans that are almost inaudible. The fans are made by attaching a tiny "patch" of piezoelectric ceramic to a metal or Mylar blade. Two factors affecting the performance of the fans are how much the ceramic patch overlaps the blade and how thick the patch is compared with the blade's thickness. Another critical factor is precisely where to attach the blade to the patch. Those factors dictate performance characteristics such as how far the blade moves, how much airflow it produces, and how that flow produces complicated circulation patterns. An improperly designed fan could actually make matters worse by recirculating hot air back onto electronic components, notes Arvind Raman, an assistant professor of mechanical engineering at Purdue.

The Purdue researchers have developed mathematical techniques that take these factors into consideration when designing fans for specific purposes. "These fans typically have been novelty items," says Raman. "If you want to really be serious about putting them into any practical use, there are so many things you need to understand about how they work and how to optimize them."

Mathematical models developed by Purdue researchers can be used to provide design guidelines for engineers. "What we bring to the table is a knowledge of the modeling of these fans," says Garimella. "How to analyze the design, to figure out how large a patch should be for how long a blade, how thick the patch should be, and what happens if you modify all these quantities. In short, it's how to optimize the performance of these fans.

Raman and his students developed relatively simple mathematical formulas that make it easier for engineers to begin designing fans for specific jobs. Engineers can use the formulas to do a quick, "back-of-the-envelope" design. "And then you might want to do some fine tuning and tweaking with more detailed analysis," says Garimella.

Chapter 9

The Critical Last Inch—Input and Output Technologies

The telecom world spends a lot of time thinking about the "last mile"—that critical distance between the customer and the service provider's equipment. Yet, for a growing number of telecom device manufacturers (and their customers), the really important factor limiting the use of telecom technology is the "last inch"—the distance that separates the user's finger from a keyboard or keypad and the user's eye from a display screen.

As telecom devices shrink, "last inch" design issues are becoming increasingly critical. For example, how do you allow people to input alphanumeric data into a button-sized PDA? And, conversely, how does one mount a screen that can display meaningful information on a device that's no larger than a thumbnail? Researchers around the world are pondering the growing input/output problem and are arriving at an array of potential solutions.

9.1 A FINGER PHONE

Mobile phone manufacturers are beginning to experiment with novel phone form factors in an attempt to better address users' daily needs. Japan's NTT DoCoMo, for instance, is developing a radically new type of mobile phone that uses the human hand as an integral part of the receiver. The FingerWhisper, which is being developed at NTT DoCoMo's Yokosuka, Japan, R&D center, works by requiring its user to stick a finger into his or her ear.

Telecosmos: The Next Great Telecom Revolution, edited by John Edwards
ISBN 0-471-65533-3   Copyright © 2005 by John Wiley & Sons, Inc.

The watch-like terminal, which is worn on the wrist, converts voice to vibration through an electromechanical actuator. It then channels the vibrations through the hand's bones to the tip of the user's index finger. By inserting this finger into the ear canal, the vibration can be heard as sound. Because the microphone is located on the inner side of the wrist, the posture of the user's hand when using the terminal is the same as when using a mobile phone.

Efforts toward developing the FingerWhisper began back in 1996, when NTT DoCoMo realized that it was approaching the limits of mobile phone miniaturization after 20 years of drastically shrinking devices. In fact, making phones smaller would require the distance between the speaker and microphone to become shorter than the actual distance between the ear and mouth, creating usability problems. With FingerWhisper, this problem is not an issue.

FingerWhisper is also designed to present an elegant solution to another key problem: providing practical input capabilities on a miniature device. The FingerWhisper eliminates the need for buttons by using an accelerometer to detect the tapping action of fingers. Combinations of finger tapping sequences serve as Morse code-like commands such as "talk" or "hang up." Through a five-stroke tapping sequence, approximately 30 commands can be issued.

NTT DoCoMo promises that FingerWhisper delivers received voices clearly even in noisy environments and allows users to speak at a lower volume compared with ordinary handsets. The watch-like design makes the unit easy to wear and, when not it use, frees the user's hands for other tasks.

The device also aims to solve a cultural problem. Earphone-microphone headsets, popular with many U.S. mobile phone users, have never caught on in Japan, perhaps became many Japanese people may not want to be seen as talking to themselves. Although no device is actually held, FingerWhisper usage conveys the impression of talking on a mobile phone and alleviates any possible sense of discomfort. On the other hand, NTT DoCoMo may have created another cultural barrier: the fact that many Western users may be reluctant to be seen sticking a finger in their ear.

In any event, NTT DoCoMo isn't the only company working on body-conduction mobile phone technology. Sanyo recently introduced the TS41 handset. This device is equipped with a "Sonic Speaker" that, when placed against its user's skull, cheekbone, or jaw, transmits sounds to the inner ear through vibrations. The product, which is now available in Japan, is designed for use in crowds and other places where external noise can drown out phone sounds.

9.2 VOICE INPUT

Certainly, the most natural way of interacting with a machine is by voice. After all, it's the primary method people use to exchange commands and information with other people. That's why researchers are working hard to develop voice recognition tools that will allow telecom device users to input information simply by speaking.

9.2.1 Saying It With Meaning

With devices growing smaller and users becoming increasingly dissatisfied with miniature keyboards and keypads, many researchers are turning their attention toward creating voice input technologies. If researchers succeed in their efforts, pressing buttons to dial phone numbers, wielding bulky remote controls, and typing on computer keyboards will all seem quaint within a decade, as effective and efficient voice input technology radically changes the ways people use telecommunications and computer products.

"We are rapidly approaching the point where entering data to devices by voice—regardless of language or accent—will be as accurate and efficient as entering it by keypad or mouse," says Lawrence R. Rabiner, associate director of the Center for Advanced Information Processing at New Jersey's Rutgers University. When this happens, another wall between humans and machines will fall. "The idea of 'going to work' to get things done will change to 'getting things done' no matter where you are," notes Rabiner, a Rutgers electrical and computer engineering professor, former vice president of research at AT&T Labs and coauthor of four books in the fields of digital signal processing and speech processing.

New technologies for compressing and transporting massive amounts of computer code, without the need for excessive network capacity, will help usher in this new age of voice control. The shrinking size of equipment will also drive the move away from hand-operated controls. "There's no room for a keypad when the device you're controlling is as small as a single key. Voice control has an advantage here because it requires virtually no physical space and we always carry our voices with us," says Rabiner. For security, new speech verification technologies will be able to analyze voices and restrict use of devices to intended users only.

Rabiner says he expects that within the next 5 to 10 years telephone calls will be made by name and not by number. Additionally, intelligent voice-controlled communications agents, essentially nonintrusive network-based robots, will place users' phone calls, track down the people that users need to reach, and let callers know whether these people are willing to talk. Voice-controlled agents will also help users find deals on merchandise, offer reminders about appointments and birthdays, and allow the control of household and office appliances from virtually any location. As a result, a wide range of home and office devices—from coffee makers to security systems—will be network accessible and voice controllable.

The distinction between work life and home life will blur as we can do whatever we want from wherever we are at any time, says Rabiner. "Work will become something we do, not someplace we go."

9.2.2 Talking to Objects

When Nassir Navab talks to inanimate objects, they usually answer him. That's because Navab, a Siemens researcher, helped develop a system that gives industrial equipment the power to vocally answer questions posed by humans.

The technology is designed to provide an easy way to check on the operational status of various gadgets, including valves, pumps, switches, and motors. Equipped with a wearable or mobile computer containing a built-in camera, a user could determine the status of any piece of equipment simply by walking around the factory floor. An 802.1-pound wireless network transfers data from the equipment to a central server and from the server to the user. A microphone-equipped headset and voice-recognition and synthesis softwares supply the user interface.

With the use of visual markers provided by the mobile camera, as well as localization software, the server automatically calculates the user's exact position and orientation. "He or she can walk around the workplace, point the camera at the object, and ask how it is," says Navab, who's an imaging and visualization project manager at Siemens Corporate Research in Princeton, New Jersey "One simply says something like, 'Current status.'" The server will then query the object and divulge the item's current operating condition— temperature, pressure, voltage, flow rate, and so on. The user can then request specific historical information about the object, such as its model number, age, service history, and the name of the employee directly responsible for its maintenance. The user can also leave a voice message that will be supplied to the next person who talks to the object.

Navab believes that the technology could be used in a wide range of fields. "Factories, power plants, refineries—any place that might have thousands of different pieces of equipment," he says. Siemens is testing the system with a variety of different wearable and mobile computers, and a pilot version is being installed at a working industrial facility.

9.2.3 Computer Talk

When using speech-based interfaces on mobile phones and other devices, people not only talk to a computer, they often find themselves talking like a computer—a finding that could be crucial for researchers designing mobile audio interfaces.

Researchers at Oregon Health & Science University in Portland, Oregon, have discovered that people who converse with text-to-speech (TTS) computer systems substantially change their speech to sound like the computer— a phenomenon known as speech convergence. Additionally, people tend to readapt their voice to each new computer-generated voice they hear.

"Human speech is always a variable," says Sharon Oviatt, an experimental psychologist at the university's OGI School of Science & Engineering. "We discovered that people subconsciously tailor their voice to the computer's voice. So, for example, if the computer voice is talking more quietly, the person interacting with the computer will talk more quietly. This is a major new source of variability in users' spoken language to computers, which is opening up new scientific directions for mobile audio interface design."

To study the phenomenon, Oviatt and her graduate students asked students ages 7 to 10 to use a special handheld computer to talk with a variety of digital marine animals. The students used the computer to speak directly to the animated software characters. The marine animals answered each question using text-to-speech (TTS) output, along with animated movement.

Four different TTS voices were used to determine whether the children accommodated their own voices—including amplitude, duration, and pitch—to more closely match the TTS output from the computer characters. Oviatt and her team are now modeling the results from these and future experiments so they'll be able to reliably predict how speech will vary under certain circumstances. Such quantitative modeling, Oviatt says, will lead to a new science of audio interface design, which will be important for developing new mobile systems that are capable of processing users' speech reliably in natural settings. "Ideally, we want to develop high-functioning applications and interface designs for specialized populations," says Oviatt. "This is a tall order for computers right now."

Humans presently must adapt to the limitations of computers, but future computers need to be more adaptive to people. "They should be smaller and embedded so we don't have to carry them," Oviatt notes. "They should be able to combine pen and speech input to express themselves naturally and efficiently and with information tailored to an individual's communication patterns, usage patterns, and physical and cognitive needs."

Oviatt's research could prove to be important to developers working to meet the demands created by a soaring speech recognition market. According to Frost & Sullivan, a technology research firm located in San Antonio, Texas, the voice recognition market will climb from $107 million in 2002 to $1.24 billion in 2009.

9.3   IMPROVED AUDIO OUTPUT

Along with developing speech recognition technologies, researchers are also working to improve audio output. New acoustic sensor research could soon help many people hear better and lead to improved audio clarity for mobile phone and other telecom devices.

The research being conducted by Ron Miles, a mechanical engineering professor at the State University of New York at Binghamton, is expected to lead to a revolution in hearing aid technology within the next four years. Miles' aim is to dramatically improve the ability of the hearing-impaired to understand speech in noisy environments. The work could help the more than 28 million Americans who already suffer from or face imminent hearing loss. The number of people with hearing problems is likely to become even larger as aging baby boomers move into their senior years. "Our focus is to improve the technology of acoustic sensing and signal processing so that we can minimize the influence of unwanted sounds," says Miles. "Research shows that hearing in noisy environments remains the number one unsolved problem faced by hearing aid wearers."

Miles' work is based on discoveries about the directional hearing capabilities of a small fly—*Ormia ochracea*. Miles has used a tiny structure found in the fly's ear as a model to develop the world's smallest directional microphones. The research holds promise in any number of civilian and military applications where microphones and acoustic-sensing systems are or could be employed.

Improving the directionality of hearing aids, enhancing their ability to filter out unwanted noise, and producing microphones that create less internal noise will mean major enhancements to speech intelligibility in noisy environments, Miles says. He notes that the improvements will be accomplished by research in three interrelated areas: directional microphones, optical electronic sensors, and signal processing. Along with Miles, the project's principal investigator, researchers Douglas Jones of the University of Illinois, an expert in signal processing algorithms, and Levent Degertekin of the Georgia Institute of Technology, an expert in optical sensors, are also participating in this research.

It is hoped that optical sensors will replace and improve on the variable capacitors used in traditional hearing-aid technology. By "reading out" sound waves hitting the microphone's diaphragm through signals created by changes in light rather than in electronic voltage, much thinner and more sensitive diaphragms can be used. "This will remove some of the key design constraints that have limited the development of small microphones," says Miles. "It should permit a revolution in microphone designs and enable the achievement of much greater sensitivity and lower noise."

The signal processing algorithms will allow for the fine-tuning and customization of hearing aid sensitivity and will reduce unwanted sounds beyond what is possible with existing hearing aid technology. Ultimately, the signal processing could be tuned based on various criteria, including directionality, frequency, or volume of sounds, Miles says. Initially, the researchers will focus on directionality, since most hearing-aid users want to hear the speaker or sound source they are facing more than other ambient room noise.

Ultimately, Miles' work will affect any application in which a miniaturized microphone and signal processing technology could improve a product's utility and performance. Besides the development of next-generation hearing aids, other envisioned applications include security devices, mobile phones, and teleconferencing equipment.

9.4   TOUCH INPUT

When words fail people, they often resort to touch. A comforting hand on the shoulder or, in dire situations, a sock to the jaw can express a person's intentions and feelings quite adequately. Touch can also be used to input various types of information on computers and many types of mobile devices.

Imagine, for example, ordering your meal in a restaurant by a simple tap on the table, which would transmit your choice directly to the kitchen. Or how about placing an order for goods by making your selection on the surface of the shop window? It may sound like science fiction, but this could be the way we interact with computers in the future, thanks to a pan-European research project, led by experts at Wales' Cardiff University.

"The vast majority of us communicate with our computers using tangible interfaces such as the keyboard, mouse, games console, or touch screen," says Ming Yang of the University's manufacturing engineering center. "Although these are in common usage, they have certain disadvantages—we are required to be 'within reach' of the computer, and most devices lack robustness to heat, pressure, and water, restricting their spheres of application. Although some voice-activated and vision systems for interacting with computers do exist, they are as yet unreliable."

The research project's goal is to develop Tangible Acoustic Interfaces for Computer Human Interactions (TAI-CHI). It will explore how physical objects such as walls, windows, and table tops can in effect become giant three-dimensional touch screens, acting as an interface between any computer and its user.

The whole project is based on the principle that interacting with any physical object produces acoustic waves both within the object and on its surface. By visualizing and characterizing such acoustic patterns and how they react when touched or moved, researchers can develop a new medium for communication with computers and the cyber-world.

Although acoustic sensing techniques have been used for many years in both military and industrial applications, none is suitable for the multimedia applications envisaged by TAI-CHI. Some commercial products also exist, but these are limited in their application to flat glass surfaces only and are restricted by size. The TAI-CHI project will go well beyond these limitations. "Our goal is to make this technology accessible to all," says Dr. Yang, who leads the TAI-CHI team. "Once that is done, the possibilities of application are endless."

9.4.1   Touching Research

As touch technology moves forward, it appears likely that it will evolve into yet another way people will communicate electronically. Right now, researchers at the University at Buffalo's Virtual Reality Laboratory are bringing a new and literal meaning to the old AT&T slogan, "Reach out and touch someone."

The engineers have created a new technology that transmits the sensation of touch over the Internet. The development could lead to the creation of technologies that teach users how to master various skills and activities, such as surgery, sculpture, playing the drums, or enhancing golf skills, all of which require the precise application of touch and movement. "As far as we know, our technology is the only way a person can communicate to another person the sense of touch he feels when he does something," says Thenkurussi Kesavadas, a University of Buffalo associate professor of mechanical and aerospace engineering and the lab's director.

Although the technology is still a long way from being able to capture and communicate the complex feel of a perfect golf swing, Kesavadas and his fellow researchers have successfully used it to transmit the sensation of touching a soft or hard object and the ability to feel the contour of particular shapes. The researchers call their technology "sympathetic haptics," which means "having the ability to feel what another person feels," says Kesavadas.

The technology communicates what another person is feeling through an active-tracking system that's connected between two PCs. The system uses a virtual reality data glove to capture the hardness or softness of an object being felt by a person. This feeling is then communicated instantaneously to another person seated at a computer terminal who, using a sensing tool, follows a point on the computer screen that tracks and transmits the movements and sensations of what the first person is feeling. "When the person receiving the sensation matches the movements of the person feeling the object, he not only understands how the person moved his hand, but at the same time he feels exactly the kind of forces the other person is feeling," explains Kesavadas.

Kesavadas notes that the sensation of touch is the brain's most effective learning mechanism—more effective than seeing or hearing—which is why the new technology holds so much promise as a teaching tool. "You could watch Tiger Woods play golf all day long and not be able to make the kind of shots he makes, but if you were able to feel the exact pressure he puts on the club when he putts, you could learn to be a better putter."

Kesavadas and his coresearchers are particularly interested in the technology's potential for medical applications. They're pursuing ways of communicating to medical students the exact pressure employed by an expert surgeon as he or she cuts through tissue with a scalpel. They also believes the technology could one day be used for medical diagnosis—allowing a doctor to feel a human organ and then to check it for injury or disease via the Internet.

A key benefit of the technology, according to Kesavadas, is its ability to capture for future replay and continual instruction the sensation of an activity after it's been transmitted. "It almost would be like one-on-one training," he says. "You could replay it over and over again. Hospitals could use it to deliver physical-therapy sessions to patients, for example."

Also researching haptics technology is the Massachusetts Institute of Technology (MIT). The school is investigating the technology's long-distance communication potential.

In 2002, MIT's technology was used in the first "transatlantic touch," which linked researchers at MIT and University College London. The experiment involved a computer and a small robotic arm that took the place of a mouse. A user could manipulate the arm by clasping its end, which resembles a thick stylus.

MIT's system creates the sensation of touch by exerting a precisely controlled force on the user's fingers. The arm, known as the PHANToM, was invented by others at MIT in the early 1990s and is available commercially through Woburn, Massachusetts-based SensAble Technologies. The researchers modified the PHANToM software for the transatlantic application.

During the demonstration, each user saw a three-dimensional room presented on a computer screen. Within the virtual room were a black box and two tiny square pointers that showed the users their position within the room. The users then used the robotic arms to collaboratively lift the box. As each user moved the arm—and therefore the pointer—to touch the box, he could "feel" the box, which had the texture of hard rubber. Together, the users attempted to pick up the box—one applying force from the left, the other from the right—and hold it as long as possible. All the while, each user could feel the other's manipulations of the box.

"Touch is the most difficult aspect of virtual environments to simulate, but we have shown in our previous work with MIT that the effort is worthwhile," says Mel Slater, professor of virtual environments in UCL's Computer Science Department. "Now we are extending the benefits of touch feedback to long-distance interaction."

"In addition to sound and vision, virtual reality programs could include touch as well," observes Mandayam A. Srinivasan, director of MIT's Touch Lab. "We really don't know all of the potential applications," he says. "Just like Bell didn't anticipate all of the applications for the telephone."

There are several technical problems that must be solved before everyday haptic applications can become available. The key problem is the long time delay, due to Internet traffic, between when one user "touches" the on-screen box and when the second user feels the resulting force. "Each user must do the task very slowly or the synchronization is lost," Srinivasan says. In that circumstance, the box vibrates both visually and to the touch, making the task much more difficult.

Srinivasan is confident, however, that the time delay can be reduced. "Even in our normal touch, there's a time delay between when you touch something and when those signals arrive in your brain," he says. "So in a sense, the brain is teleoperating through the hand." A one-way trip from hand to brain takes about 30 milliseconds; that same trip from MIT to London takes 150–200 milliseconds, depending on network traffic. "If the Internet time delays are reduced to values less than the time delay between the brain and hand, I would expect that the Internet task would feel very natural," Srinivasan says. Although improving network speeds is the researchers' main hurdle, they also hope to improve the robotic arm and its capabilities, as well as the algorithms that allow the user to "feel" via computer.

9.5 PROJECTION KEYBOARDS

Current-generation mobile phones and PDAs are limited by their data input technologies. Keyboards still take up too much physical space and handwriting recognition is time consuming and clumsy.

Canesta, a San Jose, Calif.-based input technology developer, aims to improve mobile device data input with its new "projection keyboard." The product creates a full-sized keyboard and mouse out of thin air via projected beams of light. "Electronic perception technology" is used to track finger movements in three dimensions as the user types on the image of a keyboard that's projected on any flat surface in front of the mobile device. No accessories are required.

The technology is designed to make mobile devices more useful for "real work" applications, such as substantive correspondence, the use of analytical tools, or tasks requiring a high degree of interactivity. With access to a full sized virtual keyboard, a mobile worker could use a mobile device to handle routine business e-mail correspondence or create a financial report without going back to the office.

The keyboard is implemented via a sensor and support components that mobile device makers can build into their products. The sensor, a module not much larger than a pea, resolves finger movements as the user types on the projected image. The user's movements are immediately transformed into "keystrokes" and then processed into a stream of serial keystroke data that are similar to the output generated by a physical keyboard. Mobile device makers can easily integrate this technology into a mobile or wireless device because the product supports standard software and hardware input interfaces.

The sensor includes two other miniature components: a pattern projector and a small infrared light source. The pattern projector, slightly larger than the sensor, uses an internal laser to project the image of a full-sized keyboard on a nearby flat surface. The light source invisibly illuminates the user's fingers, as he or she types on the projected surface, which could be a desk, tray table, or briefcase.

"Increasing miniaturization of devices such as mobile phones and PDAs means that efficient data entry has now become a serious design consideration," says Nazim Kareemi, Canesta's president and CEO. "Existing . . . solutions can be awkward and slow to use. Canesta's electronic perception technology offers a viable alternative by presenting the user with a familiar keyboard that can be used on any flat surface."

Another virtual keyboard proponent is Phoenix-based iBIZ Technology Corp., which offers the virtual laser keyboard. The VKB attaches to handhelds and projects the image of a full-size keyboard onto the surface where the handheld is placed, allowing the user to input text without a physical keyboard.

"There are no mechanical moving parts whatsoever in the virtual laser keyboard," says Ken Schilling, iBIZ's president and CEO. "It provides a projected image that is the perfect portable input device for PDAs. It's similar in responsiveness to regular keyboards, but extremely futuristic looking." The iBIZ virtual laser keyboard is compatible with Palm, Pocket PCs, laptops, and desktop PCs.

9.6 THOUGHT INPUT

The ultimate hands-free input technology may have been invented by Georgia State University researchers. The team has developed a Web browser that allows people to surf just by thinking.

Previous research has shown that it is possible to move a cursor by controlling neural activity. The researchers' BrainBrowser Internet software is designed to work with the limited mouse movements that neural control allows. The browser window is divided into an upper section that resembles a traditional browser and a lower control section. Common controls like "Home", "Refresh", "Print," and "Back" are grouped in the left-hand corner and provide feedback. When a user focuses his attention on a button, it becomes highlighted, and when the user successfully focuses on clicking the button, it emits a low tone. The right side of the control section displays links contained in the current Web page. This allows the user to more easily scan and click the links.

The researchers are working on a virtual keyboard with word prediction technology that will allow users to enter URLs.

9.7 OUTPUT

Output technology is on the other side of the human-device interface equation. Until about a decade or so ago, electronic display technology went almost unchanged for nearly a half century. The cathode ray tube (CRT), bulky and power ravenous, was the most widely used electronic display technology, widely used on TVs and computers. Liquid crystal displays were available on laptop computers and PDAs, but these were far too expensive for use on the desktop.

Today, CRTs are on the way out—models with CRTs are heavily discounted, and the technology appears to have reached the end of the road. Color LCD displays are cheap and abundant, and plasma displays are becoming increasingly popular for big-screen applications. In addition, researchers are working on a new wave of output devices that promise to give users better and brighter images in significantly less space.

9.8 A NEW VIEW

Tiny screens may actually be okay on mobile devices, as long as the screen can be placed close to the user's eye. "I only have a little screen, and how much bandwidth do I need to put all these pictures on a tiny screen?" asks Wim Sweldens, vice president of computing science research at Bell Labs in Murray Hill, New Jersey. The answer, he says, is to wear the display. "You take that tiny little screen and put it right next to your eye—it looks like a really big screen." That would mean a display that's integrated into glasses or perhaps even into contact lenses. "You can put a lot of data on there at very high resolution," Sweldens.

Many people, however, will balk at the idea of wearing displays, no matter how convenient the technology may be. That's why much display research is focusing on developing improved conventional-sized screen technologies.

9.9 PAPER-LIKE VIDEO DISPLAYS

Researchers at the University of Rochester and elsewhere are racing to develop a technology that would not only make flexible, paper-like video displays a reality but could make them in full color.

Companies around the world are working on doing away with bulky computer monitors and laptop displays. Marrying the versatility of a video screen with the convenience and familiarity of paper could yield a TV that you could fold into your pocket, a computer you could write on like an ordinary piece of paper, or a newspaper that can update itself. The technology being developed at the University of Rochester is based on polymer cholesteric liquid crystal (pCLC) particles, also known as "flakes," which are dispersed in a liquid host medium. These flakes in many respects resemble the metallic particles or "glitter" that are used as pigments for automobile body finishes and decorative applications and that come in a variety of colors spanning the visible and near-infrared spectrum. Unlike the more conventional particles, the apparent color of pCLC flakes can be made to change or completely disappear as they rotate in an electric field. This rotation, or "switching," effect is the underlying principle for using pCLC flakes as the active element in image displays and other applications. The flakes do not need the backlight used in typical computer screens because they reflect light the way a piece of paper does; thus a display that uses these flakes would use less electricity and could be easily viewed anywhere that a regular paper page can be read.

There are endless possibilities for surfaces that could be coated with "switchable" pCLC flakes, for example, for use in continuously changing banners in a store window or as a rewriteable paper that also accepts computer downloads. Other ideas include camouflage for vehicles that changes with the terrain, switchable solar reflectors, or filters for instruments used in fiber-optic applications and telecommunications.

The flake technology has some unique advantages when compared with other display technologies. For example, pCLC flakes are highly resistant to temperature variations, allowing them to be used in a much wider climate range than conventional liquid crystal displays. Because only a very small amount of flake motion is required to produce a relatively large effect, the response time of pCLC flakes can be competitive with the standard liquid crystal displays found in today's laptop computers, palmtop computers, and other competing electronic paper technologies.

Several different electronic paper technologies are under development in various laboratories around the world, and some are close to commercialization. Many can offer gray-scale displays, but all have had difficulties producing color. This is where pCLC flake technology has a distinct advantage given the plethora of colorful flakes available. Additionally, whereas typical electronic paper technologies use absorption to produce color and reflect light by scattering, the color produced by pCLC flakes is based entirely on reflection and is inherently polarized. This unique capability of pCLC flakes is due to their liquid crystalline properties, thus greatly broadening the scope of application to other areas beyond information and image displays.

"The ability to actively manipulate polarized light by means of an electric field is extremely useful for a large number of applications in optical technology, including switchable and tunable color filters, optical switches for fiber optics or telecommunications, and switchable micropolarizers, in addition to information displays," says research engineer Kenneth L. Marshall, who heads the team developing the technology at the Laboratory for Laser Energetics at the University of Rochester. "The ability to produce this electrically switchable polarization sensitivity in a material that can be conformally coated on flat or curved surfaces is one of the most unique and exciting aspects of this technology." Marshall sees other applications such as "smart windows" that could change color, reflect sunlight, or become completely opaque at will, environmentally robust switchable "paints," and even "patterned" particles for storage of encoded and encrypted information and document security. A more "off-the-wall" application includes living room wallpaper that one can tune to different colors or even to new patterns that have been downloaded from the Internet.

But don't expect to be finding these switchable pCLC flakes in products very soon. "There are a number of issues that need be solved first, like getting all of the flakes to move in the same direction at the same time," says Tanya Kosc, a University of Rochester doctoral candidate. "We don't have complete control over flake motion yet. Sometimes a flake will flip completely over instead of stopping at the point in its rotation that we want it to."

The size and shape of flakes largely influence how they move, but making uniform flakes is a difficult task. The initial method for producing flakes required melting the pCLC material and spreading it out at high temperature with a knife edge to form a thin layer or film. This motion helped to align the pCLC molecules uniformly to produce the bright reflection of a given color. The film was then fractured into tiny, randomly shaped flake-like particles by pouring liquid nitrogen over it.

Now, the team can create square, regularly sized flakes by molding the material through the openings of a wire mesh. They hope to make the fabrication process more efficient and to measure the behavior of the new flakes in an electric field to understand the best ways to manipulate them. "When we finally are able to make them behave uniformly," says Kosc, "we'll be able to think more about applications in actual devices."

Understanding exactly how and why an electric field causes each flake to reorient is crucial to creating a system that can be used reliably for products that can camouflage a vehicle, store encoded information, or bring the latest news in full color to a folded flexible film in your pocket.

9.9.1 Electronic Paper Display for Mobile Phones

A paper-like video display that's fast enough to present video content could allow the development of smaller, less power-hungry mobile phones, PDAs, and related devices with brighter displays.

The new display technology, developed by Philips researchers, utilizes electrowetting, a technique that's based on controlling the shape of a confined water/oil interface with an applied voltage. With no voltage applied, the colored oil forms a flat film between the water and a water-repellent coating, resulting in a colored pixel. When a voltage is applied between the coating and the water, the tension between the water and the coating changes. The tension change forces the water to move the oil aside, resulting in a partly transparent pixel or, if a reflective white surface is used, a white pixel.

Displays based on electrowetting have several important attributes. The switching between white and colored reflection is fast enough to display full-motion video content. Also, because electrowetting is a low-power and low-voltage technology, displays based on the technique can be made flat and thin and placed inside mobile devices with limited battery capacities. Additionally, electrowetting's reflectivity and contrast levels are better or equal to the output of other reflective display types. In fact, the technique's viewability is close to that of paper.

Electrowetting allows the creation of displays that are four times brighter than reflective LCDs and twice as bright as other emerging display technologies. The technique also provides a display in which a single subpixel is able to switch two different colors independently. This results in the availability of two-thirds of the display area to reflect light in any desired color. This ability is achieved by building up a pixel with a stack of two independently controllable colored oil films plus a color filter. The colors used are cyan, magenta, and yellow, a so-called subtractive system that's comparable to the principle used in ink-jet printing. Unlike LCDs, no polarizers are required, further enhancing the display's brightness.

Electrowetting is particularly well suited for applications that require a high-brightness and contrast-rich reflective display. The technology could lead, for example, to mobile phones that fold as flat as a credit card and can be slipped into a wallet. Displays could also be literally pasted onto an automobile's dashboard. It's also conceivable that mobile phone and PDA displays might be woven into clothing, allowing users to view text, photos, and videos on their sleeves.

9.9.2 Ogling OLEDs

Organic light-emitting devices (OLEDs) promise to revolutionize both desktop and mobile systems by offering ultra-thin, bright, and colorful displays without the need for space- and power-consuming backlighting. Based on layers of organic molecules that are sandwiched together, OLEDs represent the most exciting development in display technology since the introduction of the LCD.

OLEDs are already popping up on a few mobile devices. Eastman Kodak, for example, has released a digital camera—the EasyShare LS633—that features an OLED preview screen. Future OLED panels could find homes on products ranging from desktop and notebook PCs to PDAs to smart phones and a wide array of office and consumer appliances. In fact, the displays are thin and light enough to be plastered onto a wall like wallpaper or even sewn into clothing.

Researchers, however, have yet to develop screens that are larger than a few inches in diameter at a price point that's even remotely competitive with LCD technology. "We're still a long way off from OLED in a laptop," says Sam Bhavnani, a mobile computing analyst at ARS, a technology research firm located in La Jolla, California. "You might start to see 10-inch or maybe a 12-inch [screen] in the beginning of 2005, at best," he says.

Meanwhile, engineers at the University of Toronto are among many researchers worldwide working on developing a flexible OLED technology. The school's engineers recently became the first Canadian team to construct a flexible organic light-emitting device (FOLED), a technology that could lay the groundwork for future generations of bendable television, computer, PDA, and mobile phone screens. "It opens up a whole new range of possibilities for the future," says Zheng-Hong Lu, a professor in the University of Toronto's Department of Materials Science and Engineering. "Imagine a room with electronic wallpaper programmed to display a series of Van Gogh paintings or a reusable electronic newspaper that could download and display the day's news and be rolled up after use."

Today's flat panel displays are made on heavy, inflexible glass that can break during transportation and installation. Lu, working with post-doctoral fellow Sijin Han and engineering science student Brian Fung, developed FOLEDs that are made on a variety of lightweight, flexible materials ranging from transparent plastic films to reflective metal foils that can bend or roll into any shape. FOLED technology could be manufactured using a low-cost, high-efficiency mass-production method, Lu says. The team, which is already commercializing some related technology, hopes a marketable device could be created within two to three years.

Future ink-jet fabrication processes will allow liquid polymers to be inexpensively sprayed onto flexible surfaces to create full-color, animated FOLED images on everything from signs to windows to wallpaper and even product boxes. "For example, cereal boxes that have an animation on the wall of the box—that's not impossible to imagine," says Raj Apre, a research scientist at the Palo Alto Research Center, the Xerox R&D subsidiary located in Palo Alto, Calif. "A simple animation is probably less than five years away," predicts Apre.

A miniature radio receiver will permit a wide range of FOLED-enhanced products to be updatable, enabling manufacturers to instantly revise advertising or product support information. Adding a tiny transmitter and antenna will allow products to support two-way communications. Although most people probably won't want to conduct a conversation via a cereal box, the technology will allow mobile phone technology to be built into a wide range of products, such as FOLED-based newspapers, wristwatches, and even clothing, moving the world a big step closer to ubiquitous communication.

9.9.3 Polymer Displays

The flat panel display business is huge and still growing rapidly. However, along with the demand for ever larger displays—for example, wall-sized TVs—comes a big price tag because big displays are still made by the same expensive photolithography techniques as the diminutive silicon chip. What is needed is a completely new manufacturing approach that will dramatically lower the cost.

Polymeric, or plastic, semiconductors provide an exciting opportunity to solve the problem. Polymers can be dissolved in a liquid, thus creating a semiconducting ink. This ink can be printed using the same technology that is used in jet printers that print documents. Printing has a low cost compared with photolithography for manufacturing of electronics because both material deposition and patterning are done simultaneously. Enormous progress has been made in recent years to develop plastic semiconductors that have electronic properties suitable to drive a display.

High-speed, reproducible, and reliable processes, such as roll-to-roll display manufacturing, are also proving effective in the fabrication of light-emitting polymers (LEPs). By using ink-jet printing and the silk screening of organic materials, LEPs are able to reduce manufacturing cost through effective material utilization.

"LEPs offer all the advantages of small-molecule technology such as low-power consumption and low-drive voltages," says Joe Constance, an analyst at Technical Insights, a technology research company located in San Antonio, Texas. "LEP devices can generate sharp light output, or resolution, and can be fabricated cost effectively in high-pixel density configurations," he adds.

Efficient control over structural order in LEPs is required to have an edge over traditional liquid crystal displays. In this regard, polymers that have different band gaps may prove to be a key factor in outplaying competing technologies. Emission of red, green, or blue light is possible with different bands, making full-color displays with conductive LEPs commercially viable. Intense research in this area has enabled poly-phenylene vinylene (PPV) to emit blue light by interrupting conjugation in the polymer with nonconjugated units. Attachment of alkoxy side groups to the phenylene rings of the PPV enables red light production.

A full-color polymer-based display requires pixellating the colors through the combination of different conducting polymers with varying band gaps. The color emitted from the blend will depend on the amount of voltage applied, which increases with the band gap. An alternative route would be to use white-light-emitting diodes to create a microcavity, the length of which determines the color of the emitted light.

Apart from varying band gaps, polymer-based light-emitting diodes also face the challenge of increasing operating lifetimes to at least 20,000 hours to popularize commercial usage. When the luminance intensity of the device decreases to 70 percent of the original value, it is considered the actual end of life as opposed to 50 percent referenced for display applications. "In the area of LEPs, significant research has been in progress to improve material lifetimes both through the use of materials that are resistant to oxidation and through improved encapsulation," explains Constance.

Conducting polymers based on doped polyaniline, conjugated polymer material, and polypyrrole are already demonstrating the stability required for commercial applications, proving the potential for growth and wider acceptance in the future.

Researchers at the Xerox Research Center of Canada recently announced a new polymer in the polythiophene family. This polymer has the best electrical properties of any reported plastic semiconductor. Scientists at Xerox's Palo Alto Research Center (PARC) succeeded in jet-printing this material and other polymer semiconductors to make transistors. The jet-printed transistors made this way have matched the performance of the same material deposited by conventional spin-coating (which gives an unpatterned film), showing that the jet-printing process does not adversely affect the performance of the device. The transistors have exceptional performance for polymers and meet all the requirements for addressing displays. Along with a high mobility, they have very low leakage and good stability.

There is much more involved in the fabrication of a low-cost transistor array than just printing the polymer semiconductor. As with any integrated electronic device, metals and insulators must also be deposited and patterned into a multi-layer structure having the right electronic circuit and an appropriate physical size. The PARC scientists have successfully integrated the jet-printed polymer into a prototype display circuit, in which printing techniques define all the patterns. The electronic properties and physical dimensions meet the needs of flat-panel displays, and the complete absence of photolithography promises low-cost manufacture. The PARC array design also solves key issues of unwanted interactions between pixels of the display, accurate layer-to-layer alignment, and materials compatibility.

Although much more development is needed to make the jet-printed organic semiconductor display process ready for manufacture, this breakthrough demonstration at PARC represents proof that it can be done successfully.

9.9.4 Quantum Displays

Massachusetts Institute of Technology (MIT) researchers have developed a new display technology that promises to someday replace conventional LCDs. The technology—a quantum dot-organic light-emitting device (QD-OLED)— allows the creation of flat-panel screens that consume less power and produce better images than existing counterparts.

Based on high-performing inorganic nanocrystals—combined with organic materials—QD-OLEDs would be ideal for use in mobile devices. Unlike traditional LCDs, which must be lit from behind, quantum dots generate their own light. Depending on their size, the dots can be "tuned" to emit any color in the rainbow. Better yet, the colors they produce are highly saturated, permitting richer, more lifelike images. Also known as "artificial atoms," quantum dots are nanometer-scale "boxes" that selectively hold or release electrons.

The latest MIT QD-OLED contains only a single layer of quantum dots sandwiched between two organic thin films. Previous QD-OLEDs used anywhere from 10 to 20 film layers. The researchers have created QD-OLEDs over a 1-cm^2 area, although the same principle could be used to make larger components.

The MIT team's method of combining organic and inorganic materials could also pave the way for new technologies and enhance understanding of the physics of these materials. Besides allowing the creation of extraordinarily thin, bright flat-panel displays, the QD-OLEDs may also be used to calibrate wavelengths for scientific purposes, generate wavelengths visible only to robot eyes or to "miniaturize scientific equipment in ways we haven't yet imagined," says Moungi Bawendi, an MIT chemistry professor.

The QD-OLEDs created in the study have a 25-fold improvement in luminescent power efficiency over previous QD-OLEDs. The researchers note that in time the devices may be made even more efficient and be able to achieve even higher color saturation. "One of the goals is to demonstrate a display that is stable, simple to produce, flat, high-resolution and that uses minimal power," says Vladimir Bulovic, an assistant professor of electrical engineering and computer science at MIT.

The MIT researchers were inspired by advances in completely OLED technology. OLEDs, which can be used to create TVs or computer screens only a fraction of an inch thick with the same brightness as LCDs, are now making their way into commercial electronic devices. The MIT group envisions that QD-OLEDs will in time become complementary to OLEDs because they can be built on the same electronic platforms with compatible manufacturing methods.

The QD-OLED research is funded by the National Science Foundation and Universal Display Corp., an OLED technology developer located in Ewing, New Jersey.

9.10 FINDING INFORMATION

Given their imperfect input/output technologies, finding information on tiny devices can be highly problematic. Researchers, however, are working on technologies that promise to make finding hard-to-access information a snap.

Researchers at the University of Southern California (USC), for instance, have created a new tool for organizing and visualizing collections of electronic mail. The system is designed to help legal researchers, historians, and archivists, as well as ordinary business people, deal with large e-mail archives.

Developed by Anton Leuski, a researcher at the USC School of Engineering's Information Sciences Institute, the eArchivarius system uses sophisticated search software developed for Internet search engines like Google to detect important relationships between messages and people. The technology takes advantage of inherent clues that exist in e-mail collections and then automatically creates a vivid and intuitive visual interface that uses spheres grouped in space to represent relationships.

In an experimental exercise, USC researchers collected e-mail exchanges conducted between Reagan administration national security officials. The visualization showed some e-mail recipients closely packed toward the display's center and their most frequent correspondents bunched into a tight cluster. Less frequent correspondents were literally out of the loop, with their spheres located far out on the display's periphery.

Besides correspondence activity, spheres representing people can also be arranged under other criteria, such as the content of the authored messages. The display configuration would then show communities of people who converse on the same topic and the relationships among those communities.

With the eArchivarius system, selecting any e-mail recipient opens a new window, which provides a list of all the people with whom the selected person exchanged correspondence. A time-graphed record also shows when the exchanges took place. "For a historian trying to understand the process by which a decision was made over a course of months, this kind of access will be extremely valuable," says Leuski.

The same interface can instantly return and display individual pieces of mail in the form of hypertext pages, with links to the people who sent and received the e-mail and with links to similar e-mail messages. "Similar messages" can be defined in terms of recipients, text keywords, or both. In a display produced using this capability, the spheres are the messages themselves.

With message-oriented spheres, different colors indicate different topics, creating a map of how the information is distributed among the messages. "What we have in effect is a four-dimensional display, with color added to the three spatial dimensions," says Douglas Oard, an associate professor of computer science from University of Maryland's College of Information Studies and its Institute for Advanced Computer Studies.

Leuski and Oard have demonstrated the ability to find interesting patterns in collections as small as a few hundred e-mails. The techniques they have developed are now being applied to thousands of e-mails sent and received by a single individual over 18 years. Scaling the system upward to process millions of e-mails involving thousands of people will be the researchers' next challenge.

The elements of eArchivarius' flexible and useful interface, says Oard, may someday find their way into everyday e-mail client software.

9.10.1 Simplified Image Retrieval

Penn State University has developed software that allows computer users to search for images more easily and accurately by eliminating the need to enter lengthy descriptive phrases. The technology could potentially help mobile device users who currently must deal with cramped displays as well as people who must routinely catalogue or access large numbers of images.

The Automatic Linguistic Indexing of Pictures (ALIP) system builds a pictorial dictionary and then uses it to associate images with keywords. The technology functions like a human expert who annotates or classifies terms. "While the prototype is in its infancy, it has demonstrated great potential for use in biomedicine by reading x-rays and CT scans, as well as in digital libraries, business, Web searches and the military," says Dr. James Z. Wang, an assistant professor at Penn State's School of Information Sciences and Technology and the project's lead researcher.

Unlike other content-based retrieval systems that compare features of visually similar images, ALIP uses verbal cues that range from simple concepts such as "flowers" and "mushrooms" to higher-level terms such as "rural" and "European." ALIP can also classify images into a larger number of categories than other systems, thereby broadening the uses of image databases.

ALIP processes images the way people do. When someone sees a new kind of vehicle with two wheels—a seat and a handlebar, for example—it's recognized as a bicycle from information about related images stored in the brain. ALIP has a similar bank of statistical models that "learn" from analyzing specific image features.

Other advantages include ALIP's ability to be trained with a relatively large number of concepts simultaneously and with images that are not necessarily visually similar. In one experiment, researchers trained ALIP with 24,000 photographs found on 600 CD-ROMs, with each CD-ROM collection assigned keywords to describe its content. After "learning" these images, the computer then automatically created a dictionary of concepts such as "building," "landscape," and "European." Statistical modeling enabled ALIP to automatically index new or unlearned images with the dictionary's linguistic terms.

Future research will be aimed at improving ALIP's accuracy and speed. ALIP's reading of a beach scene with sailboats yielded the keyword annotations of "ocean," "paradise," "San Diego," "Thailand," "beach" and "fish." Although the computer was intelligent enough to recognize the high-level concept of "paradise," additional research will focus on making the technology more accurate, so that San Diego and Thailand will not appear in the annotation of the same picture, Wang says. "This system has the potential to change how we handle images in our daily life by giving us better and more access," says Wang.

9.11 DISABLED ACCESS

Disabled individuals are typically forgotten in discussions about input/output technologies. Mobile phones, for instance, are still a work in progress with regard to meeting the needs of individuals with disabilities, who are missing out on wireless communication opportunities because of usability problems.

9.11.1 Mobile Phone Interface

Virginia Tech's Tonya Smith-Jackson, assistant professor, and Maury Nussbaum, associate professor, both in the Grado Department of Industrial and Systems Engineering, are conducting research to improve the cell phone interface for users with disabilities. The Telecommunications Act of 1996 placed the demand on manufacturers of mobile phones to support accessibility for individuals with physical, visual, or cognitive disabilities. "Users with disabilities have been systematically marginalized in the information age because manufacturers and designers have either ignored their needs altogether or designed features in a haphazard manner that were intended to enhance accessibility, yet resulted in unusable products," says Smith-Jackson.

While trying to operate a mobile phone, users with disabilities typically encounter problems such as small and flat buttons that are difficult to push with bent fingers, keypads with no Braille or tactile feedback to assist with orientation, or lack of voice activation capabilities. Sometimes, special features are available for disabled customers, but the features do not perform consistently, such as voice-activated phones failing to work in a noisy environment. People who have more than one type of disability have even greater difficulty operating mobile phones.

The first goal of this research is to identify user requirements and challenges related to user interface designs of cell phones. The second research goal is to conduct usability tests with existing interfaces of selected Toshiba phones designed for the Japanese domestic market that will be marketed in the United States. As part of the study for Toshiba, the researchers and their graduate students are using product interactive focus groups and usability testing to target the needs of users with the following disabilities: legal blindness, cognitive disabilities, full blindness, and upper extremity physical disabilities. Information from these interviews is being used to extract design guidelines to enhance cell phone accessibility and to develop new features for future cell phone interfaces.

9.11.2 GPS Guidance

Telecom technologies, when equipped with special interfaces, can also help disabled people cope better with the real world. A new GPS-based navigation developed by ONCE, the Organization of Spanish Blind people, is designed to guide blind people. The system, called "Tormes," is a handheld computer with a Braille keyboard and satellite navigation technology that gives verbal directions. Tormes can be used in two ways: to guide the user to their destination or to tell them where they are as they walk around.

Tormes is currently limited by GPS's 15 to 20 meter accuracy. But ONCE and the European Space Agency are already working on how to improve the system. A new tool developed •• by ESA could be the breakthrough: the European Geostationary Navigation Overlay Service (EGNOS). It also warns the users of any problem with the signal thus giving integrity information.

EGNOS is transmitted to the ground via geostationary satellites, so signals are sometimes blocked by buildings, called the canyon effect. To solve this problem, ESA engineers had the idea of getting the data through the Internet via a GSM connection, a project called SISNeT (Signal In Space through Internet). This makes EGNOS available anywhere downtown. As a result, blind people accessing information via Tormes will be able to distinguish individual streets as they approach them.

Ruben Dominguez, a blind mathematician who has tried out the device, says, "This completes what exists for assisting blind people: the dog or the white cane, but furthermore it will really improve the life of the blind community by giving a lot more autonomy when moving around town, especially in unknown places." EGNOS is scheduled to become operational by 2005.

9.11.3 Speech-Controlled Arm

Using two motors, speech-recognition software, and an exoskeleton inspired by science fiction, three Johns Hopkins University undergraduates have designed and built a muscle-enhancement device specifically for a disabled person that will help him lift a cup, a book, and other household items. By uttering commands such as "open" and "raise," this user will receive mechanical help in moving his fingers and bending his elbow. The motorized plastic shell will fit over the right arm of the man, who has an extremely rare degenerative muscle disorder called inclusion body myositis.

This device, which could be adapted for other people with disabilities, was developed by students in the Department of Mechanical Engineering's Senior Design Project course. The project originated when the man with the muscle disease sought help from Volunteers for Medical Engineering, a nonprofit Baltimore group that uses technology to assist people with disabilities. The client explained that his nerves were intact, meaning that he could control the placement of his fingers around an object. But progressive muscle deterioration left him unable to grasp and lift even small objects.

To help him, the Volunteers for Medical Engineering sponsored a project in the Johns Hopkins course. The task of designing and building the device went to a team consisting of three senior students: Jonathan Hofeller, a mechanical engineering student; Christina Peace, a biomedical engineering student; and Nathaniel Young; a biomedical engineering student. The students researched prosthetic limbs, and, taking a cue from props featured in the film "Aliens," they designed a plastic exoskeleton that could slide over the client's right hand and arm. To help move his fingers and elbow, the students tested and rejected systems using electromagnets and air pressure systems. They finally settled on two small but powerful stepper motors. These could move the fingers and elbow in small, slow increments, allowing the client to clasp a cup firmly without crushing it. In addition, these motors did not require continuous electrical current to stay in position, which preserves battery power. The students linked the motors to a series of cables and springs to enable the device to move the man's arm into position and help his fingers grasp and release.

The students opted for voice recognition software as an easy way for the disabled man to control the grasping device. After the software is trained to the client's voice, the man will first say "arm" or "hand" to take command of one of the two motors. The elbow motor will then respond to "raise," "down," or "stop." The hand motor will respond to "open," close," and "stop." The device is hard-wired to a control box that contains a miniature computer and two programs that turn the voice commands into signals that tell the motors how to operate the bending and grasping motions. The unit is powered by a rechargeable 12-volt lead-acid battery commonly used for remote-control model boats and airplanes. The control box fits inside a small pack that the man can carry on his waist, making the grasping the device fully portable.

"[The students] came up with a very creative design for the device," says Jan Hoffberger, executive director of Volunteers for Medical Engineering. "They purposely set it up to move very slowly, so that at any time in the grasping and lifting process, our client can tell it to stop. We believe he will find it very helpful."

The students had to work within a budget of $10,000; they ended up spending about $8,000 on the device. Designing and building it helped the undergraduates to understand some of the challenges that working engineers face. "In a textbook, there is always one right answer," says Young. "In this project, there were many different ways we could go, but once we were committed we had to go in that direction." His teammate, Hofeller, says, "The project involved a lot of trial and error, but it was fun to apply what we've been learning." The third team member, Peace, added, "When you're working out a problem in an engineering book, the conditions are ideal. In this project, the conditions were not perfect, but we still got the job done."

Glossary

1G: – *See* First-generation services.

2G: – *See* Second-generation services.

3G: – *See* Third-generation services.

4G: – *See* Fourth-generation services.

802.11x: – A series of IEEE standards for wireless LANs, including 80211.a, 80211.b, and numerous others.

Algorithm: – A step-by-step mathematical procedure for solving a problem.

Anechoic: – Free from echoes and reverberations.

Asset tracking: – Technology that is used to follow the physical movements of objects and people.

Baby Bell: – One of the original Bell System operating companies.

Bandgap: – The energy difference in a material between its nonconductive state and its conductive state.

Bell System: – Refers to AT&T and its Bell operating companies, which dominated the U.S. telephone industry until a court-ordered breakup in 1984.

Bit: – The smallest element of computer information.

Bits per second (bps): – A data network speed measurement. A 10-Mbps, network, for example, has a top data transfer speed of 10 million bits per second.

Telecosmos: The Next Great Telecom Revolution, edited by John Edwards
ISBN 0-471-65533-3   Copyright © 2005 by John Wiley & Sons, Inc.

Blackberry: – A line of wireless e-mail devices produced by Research In Motion.

Bluetooth: – An open standard for the short-range transmission of digital voice and data between mobile devices.

BPL: – *See* Broadband over power lines.

bps: – *See* Bits per second.

Broadband: – High-speed Internet access, faster than 56K bps dial-up service.

Broadband over power lines (BPL): – An Internet access technology that use poker lines.

Buckyball: – A spherical carbon molecule, also known as a "Fullerine," composed of 60 atoms. Buckyballs are lighter than plastic and stronger than steel.

Cable modem: – A device that connects a computer to a cable television system's broadband Internet service.

Carrier: – A telecommunications service provider.

Cathode ray tube (CRT): – A vacuum tube that serves as a computer display.

CDMA: – *See* Code division multiple access.

CDMA 2000: – *See* Code division multiple access 2000.

CLEC: – *See* Competitive local exchange carrier.

Code division multiple access (CDMA): – A second-generation (2G) digital mobile phone technology that operates in the 800-MHz and 1.9-GHz PCS bands.

Code division multiple access 2000 (CDMA 2000): – A CDMA version for third-generation (3G) networks.

Competitive local exchange carrier (CLEC): – A local telephone carrier that was not one of the original Bell System operating companies.

Constellation: – An array of satellites that is designed to provide continuous, or near-continuous, access from any point on earth.

CRT: – *See* Cathode ray tube.

Cybersecurity: – The protection of computers and networks.

DARPA: – *See* Defense Advanced Research Projects Agency.

Data hiding: – *See* Steganography.

Defense Advanced Research Projects Agency (DARPA): – The central research and development organization for the U.S. Department of Defense (DoD).

Dense wavelength division multiplexing (DWDM): – A higher capacity form of wavelength division multiplexing.

Dielectric: – An insulator, such as glass or plastic.

Digital radio: – A radio based on digital technology.

Digital subscriber line (DSL): – High-speed Internet access service using phone lines.

Downlink: – A communications channel that sends audio and/or video from a satellite to earth.

Downstream: – A communications channel that sends data from a satellite to earth.

DSL: – *See* Digital subscriber line.

DWDM: – *See* Dense wavelength division multiplexing.

EDGE: – *See* Enhanced data rates for global evolution.

Encryption: – The process of transforming plain information into a secure format, designed to protect its confidentiality.

Enhanced Data Rates for Global Evolution (EDGE): – An enhancement to TDMA and GSM that boosts data speeds to 384,000 bits per second.

ESA: – *See* European Space Agency.

European Space Agency (ESA): – The organization that manages the European space program on behalf of 15 member states.

Extensible Markup Language (XML): – An open standard for describing data that's used for defining data elements on a Web page and business-to-business documents. XML has become the standard for defining data interchange formats on the Internet.

4G: – *See* Fourth-generation services.

Fiber: – *See* Optical Fiber.

First-generation services (1G): – Analog mobile phone services.

Flat panel: – A thin display that use LCD, plasma, or other type of non-CRT technology.

Fourth-generation services (4G): – Ultra-high-speed multimedia digital mobile phone services.

Fractal: – An object that is self-similar at all scales, in which the final level of detail is never reached and never can be reached by increasing the scale at which observations are made.

Fuel cell: – A device that converts a gas or liquid fuel into electricity to power a notebook computer, mobile phone, or other electronic product.

Galileo: – A satellite-based radio navigation system currently under construction by the European Space Administration (ESA).

GHz: – *See* Gigahertz.

GIF: – *See* Graphics interchange format.

Gigahertz (GHz): – One billion cycles per second. *See* Hertz.

Global positioning system (GPS): – A satellite-based radio navigation system that allows users to find their precise location anywhere on earth.

Global system for mobile communications (GSM): – A second-generation (2G) digital mobile phone technology based on TDMA that is the predominant system in Europe and is gaining increasing popularity in North America.

GPS: – *See* Global positioning system.

Graphical user interface (GUI): – A graphics-based user interface that features windows, icons and pointing device input.

Graphics interchange format: – A popular graphics file format.

GSM: – *See* Global System for Mobile Communications.

GUI: – *See* Graphical User Interface.

Haptic interface: – Communicating with a computer via touch sensation.

Hertz (Hz): – The basic unit of electrical cycles.

Hotpsot: – A place, such as a home or store, where a wireless connection is available.

IEEE: – *See* Institute of Electrical and Electronics Engineers.

ILEC: – *See* Incumbent Local Exchange Carrier.

IP Telephony: – The two-way transmission of audio over a network that uses Internet protocols.

IM: – *See* Instant messaging.

Incumbent local exchange carrier (ILEC): – A local telephone carrier that was one of the original Bell System operating companies.

Information hiding: – *See* Data Hiding.

Instant messaging (IM): – The process of exchanging real time voice or text messages over a network.

Institute of Electrical and Electronics Engineers (IEEE): – A membership organization that sets many telecommunications, networking, and computer standards.

Interference: – Unwanted signals from a manmade or natural source.

International Organization for Standardization (ISO): – An international standards-setting organization.

Internet protocol (IP): – The network layer protocol in Internet-based networks.

IP: – *See* Internet Protocol.

ISO: – *See* International Organization for Standardization.

Joint Photographic Experts Group: – The organization that developed JPEG, a popular image file compression format.

JPEG: – A popular image file compression format. *See* Joint Photographic Experts Group.

Key: – In security, a numeric code that's used to encrypt information.

kHZ: – *See* Kilohertz.

Kilohertz (kHz): – One thousand cycles per second. *See* Hertz.

LAN: – *See* Local area network.

Local area network (LAN): – A computer network that serves users in a confined location, such as an office or building.

Local loop: – The connection between a phone customer and the phone company's office.

Location-based service: – A service that works by pinpointing its user's location.

Megahertz (MHz): – One million cycles per second. *See* Hertz.

MEMS: – *See* Micro-electrical mechanical systems.

Mesh network: – A network that provides at least two pathways between each node.

MHz: – See Megahertz.

Micro-electrical mechanical systems (MEMS): – Nano-sized devices that are built onto chips.

Microscillator: – A miniature device for generating tunable microwave signals.

Motion Pictures Experts Group: – The organization that developed MPEG, a popular video compression format.

Motion tracking: – Using a video system to automatically follow a moving person or object.

MP3: – A popular audio compression format.

MPEG: – A popular video compression format. *See* Motion Pictures Experts Group.

MSO: – *See* Multiple System Operator.

Multiple System Operator (MSO): – A cable TV company or other organization that has franchises in various locations.

Nanotechnology: – The creation of materials and devices at atomic and molecular levels.

Nanotube: – A carbon molecule, resembling a chicken wire cylinder, that's approximately a millimeter long and about one to two nanometers in diameter. Featuring a tensile strength 10 times greater than steel at about one-quarter the weight, nanotubes are considered the strongest known material for their weight.

NASA: – *See* National Aeronautics and Space Administration.

National Aeronautics and Space Administration (NASA): – The U.S. government agency that operates the nation's space program.

National Institute of Standards and Technology (NIST): – A U.S. government agency that develops and promotes measurements, standards, and technologies to enhance productivity, facilitate trade and improve the quality of life.

National Science Foundation (NSF): – An independent US government agency responsible for promoting **science** and engineering.

NIST: – *See* National Institute of Standards and Technology.

Node: – In a network, a computer, printer, hub, router or other connection or interconnection point.

NSF: – *See* National Science Foundation.

OLED: – *See* Organic light-emitting diode.

Omnidirectional: – Describes a device, such as a microphone or antenna, that emits or receives signals from all directions.

Optical Fiber: – A thin glass strand designed to carry voice or data signals.

Organic light-emitting diode (OLED): – A technology that provides ultra-thin, bright, and colorful displays without the need for space-hogging and power-consuming backlighting.

Oxygen: – *See* Project Oxygen.

PAN: – *See* Personal area network.

Passband: – A spectrum segment that is allowed to pass between two limiting frequencies.

Personal area network (PAN): – A short-range network, usually wireless, that provides a connection between two or more devices, for example, linking a PDA to a computer in order to synchronize data.

Photon: – A particle of light.

Photonic circuit: – A circuit that uses light rather than electricity.

Photonic crystal: – A credit-card-thick stack of optical filters.

Picture phone: AT&T's video telephone technology, introduced at the 1964–1965 New York World's Fair.

Piezoelectric: – Material that moves when placed under an electric voltage.

Plain old telephone service (POTS): – Ordinary telephone lines and equipment.

Polymer: – A substance made of repeating chemical units or molecules. The term is often used in place of plastic or rubber.

POTS: – *See* Plain old telephone service. Ordinary telephone lines and equipment.

Project Oxygen: – A Massachusetts Institute of Technology project for replacing discrete telecommunications and computer devices with a ubiquitous—often invisible—infrastructure.

Protocol: – Rules pertaining to the transmission and reception of information.

Quantum cryptography: – Technology for encrypting data that draws on inherent properties of photons.

Radio frequency identification (RFID): – An asset tracking and data collection technology that uses electronic tags to store identification data and a remote reader to capture information.

RBOCs: – *See* Regional Bell operating companies.

Reader: – A device that obtains data from a source, optically, electrically, or via radio or infrared signals.

Regional Bell operating companies (RBOCs): – One of the original Bell System operating companies.

RFID: – *See* Radio frequency identification.

SALT: – *See* Speech application language tags.

SDR: – *See* Software-defined radio.

Second-generation services (2G): – Digital mobile phone services.

Sensor: – A device that detects a real-world condition, such as heat, motion or light, and converts and relays that information to a computer.

Smart: – Intelligence built into a device or system.

Smart phone: – A telephone with telecommunications, information access and data processing capabilities.

Smart appliance: – A household appliance with Internet connectivity.

Software-defined radio (SDR): – A radio that can be instantly adapted to accommodate any standard, simply by loading in various programs.

Speech application language tags (SALT): – A document language format code that makes speech applications accessible from GUI-based devices, such as PCs and PDAs.

Speech integration: – Technology that adds voice services to enterprise phone systems and Web sites.

Speech recognition: – *See* Voice recognition.

Steganography: – An encryption technique for hiding a message inside an image, audio, or video file.

Surveillance: – The observation of an area or of people or objects.

Tag: – An RFID device that contains information about a particular asset. Also a document language format code.

TDMA: – *See* Time division multiple access.

Teleconference: – An audio or audio/video conference of geographically dispersed people using a telecommunications network.

Telehomecare: – The practice of using medical devices to relay medical information to a caregiver on behalf of a home-based patient.

Telematics: – *See* Vehicular telematics.

Telemedicine: – Health care practiced over distance by a network connection.

Testbed: – An environment used to test a specific project.

Thin Film: – A thin layer of material that's deposited onto a metal, ceramic or semiconductor base.

Third-generation services (3G): – High-speed multimedia digital mobile phone services.

Time division multiple access (TDMA): – A second-generation (2G) mobile phone technology that interleaves multiple digital signals onto a single high-speed channel.

Ultra wideband radio (UWB): – A radio that uses ultra-short pulses to distribute power over a wide portion of the radio frequency spectrum.

Because power density is dispersed widely, UWB transmissions ideally won't interfere with the signals on narrow-band frequencies.

Uplink: – A communications channel that sends audio and/or video from earth to a satellite.

Upstream: – A communications channel that sends data from earth to a satellite.

UWB: – *See* Ultra wideband radio.

VCSEL: – *See* Vertical Cavity Surface Emitting Laser.

Vehicular telematics: – Vehicle-based information, entertainment, and navigation systems.

Vertical Cavity Surface Emitting Laser (VCSEL): – A laser diode that emits light from its surface rather than its edge.

Voice over Internet protocol (VoIP): – A form of IP telephony that allows people to place telephone calls over Internet connections.

Voice recognition: – The conversion of spoken words into computer-usable data.

VoiceXML: – An XML extension for creating telephone-based, speech-user interfaces.

VoIP: – *See* Voice over Internet protocol

WAN: – *See* Wide area network.

Waveguide: – A device for confining and directing electromagnetic waves.

Wavelength division multiplexing (WDWM): – A technology that utilizes multiple lasers to send several wavelengths of light simultaneously over a single optical fiber. Each signal travels within a separate color band.

WDWM: – *See* Wavelength division multiplexing.

Wearable computer: – A computer that can be attached to its user's body or worn as a garment.

Webconference: – A text, audio or text/audio/video conference of geographically dispersed people using the Internet's World Wide Web.

Web services: – Software that knows how to talk to other types of software over a network. A Web service can be nearly any type of application that has the ability to define to other applications what it does and can perform that action for authorized applications or parties.

Wide area network (WAN): – A computer network that serves users in multiple locations, may be regional, nationwide or even global in scope.

Wi-Fi: – A certification for 802.11 wireless network products that comply with Wi-Fi Alliance specifications. Also used as a slang term for 802.11 wireless network products in general.

Wi-Fi Alliance: – A trade organization of 802.11 wireless network product vendors.

Wireless local area network (WLAN): – A local area network that uses a radio technology, such as 80211.x, to interconnect nodes.

Wireline: – Telephone service provided by wire or cable, as opposed to mobile phone service.

WLAN: – *See* Wireless Local Area Network.

WPAN: – *See* Personal Area Network.

XML: – *See* Extensible Markup Language.

ZigBee: – A standards-based wireless networking technology that supports low data rates, low power consumption, security, and reliability. ZigBee is designed to address the unique needs of most remote monitoring network applications.

Index

Telecosmos: The Next Great Telecom Revolution, edited by John Edwards
ISBN 0-471-65533-3 Copyright © 2005 by John Wiley & Sons, Inc.